Small Firms and U.S. Technology Policy

For Carol

For Amy, Jasper, and Alex

Small Firms and U.S. Technology Policy

Social Benefits of the U.S. Small Business Innovation Research Program

Albert N. Link

Virginia Batte Phillips Distinguished Professor of Economics, Department of Economics, University of North Carolina at Greensboro, USA

Martijn van Hasselt

Associate Professor of Economics, Department of Economics, University of North Carolina at Greensboro, USA

 Edward Elgar
PUBLISHING

Cheltenham, UK • Northampton, MA, USA

Published by
Edward Elgar Publishing Limited
The Lypiatts
15 Lansdown Road
Cheltenham
Glos GL50 2JA
UK

Edward Elgar Publishing, Inc.
William Pratt House
9 Dewey Court
Northampton
Massachusetts 01060
USA

A catalogue record for this book
is available from the British Library

Library of Congress Control Number: 2023930361

This book is available electronically in the **Elgar**online
Business subject collection
http://dx.doi.org/10.4337/9781802205596

ISBN 978 1 80220 558 9 (cased)
ISBN 978 1 80220 559 6 (eBook)

Printed and bound in Great Britain by TJ Books Limited, Padstow, Cornwall

Contents

List of figures   vii
List of tables   x
About the authors   xv
Acknowledgments   xvii
List of abbreviations   xviii

PART I   SBIR PROGRAM: INSTITUTIONAL PERSPECTIVES

1   Introduction to *Small Firms and U.S. Technology Policy*   2

2   Legislative history of the SBIR program   8

3   The SBIR program: an element of U.S. technology policy   22

PART II   SBIR PROGRAM: EMPIRICAL PERSPECTIVES

4   SBIR databases and empirical studies of the SBIR program   30

5   Demographics of SBIR awardees   39

6   Productive capital   44

7   When research fails   54

8   University collaborators on SBIR projects   70

9   Knowledge production functions   78

10   Financial stakeholders in SBIR projects   88

11   Market for SBIR developed technologies   94

PART III   SBIR PROGRAM: POLICY PERSPECTIVES

12   Unanticipated consequences   99

13   A counterfactual analysis   112

14   Concluding remarks   121

References   124
Index   130

Figures

1.1	The organization of the book	3
2.1	Multifactor productivity index for the U.S. private business sector (2012 = 100)	8
2.2	Year-to-year percentage change in multifactor productivity index for the U.S. private business sector	9
2.3	Risk and expected rate of return for a private-sector firm as a result of SBIR funding	17
3.1	Representation of U.S. technology policy	22
3.2	The policy transfer of the SBIR program to other countries	27
5.1	DOD Phase II awards to women-owned and minority-owned firms, by award year	40
5.2	NIH Phase II awards to women-owned and minority-owned firms, by award year	41
5.3	NASA Phase II awards to women-owned and minority-owned firms, by award year	41
5.4	DOE Phase II awards to women-owned and minority-owned firms, by award year	42
5.5	NSF Phase II awards to women-owned and minority-owned firms, by award year	42
7.1	Percent of Phase II DOD-funded projects that failed, by award year, n = 1,463	55
7.2	Percent of Phase II NIH-funded projects that failed, by award year, n = 820	56

7.3   Percent of Phase II NASA-funded projects that
        failed, by award year, n = 306   56

7.4   Percent of Phase II DOE-funded projects that
        failed, by award year, n = 315   57

7.5   Percent of Phase II NSF-funded projects that
        failed, by award year, n = 316   57

7.6   Percent of DOD Phase II projects that failed, by
        primary reason for being discontinued and by
        award year, n = 1,463   65

7.7   Percent of NIH Phase II projects that failed, by
        primary reason for being discontinued and by
        award year, n = 820   66

7.8   Percent of NASA Phase II projects that failed,
        by primary reason for being discontinued and by
        award year, n = 306   67

7.9   Percent of DOE Phase II projects that failed, by
        primary reason for being discontinued and by
        award year, n = 315   68

7.10   Percent of NSF Phase II projects that failed, by
        primary reason for being discontinued and by
        award year, n = 316   69

8.1   Percent of Phase II DOD-funded projects that
        involved university resources, by award year, n = 1,463   72

8.2   Percent of Phase II NIH-funded projects that
        involved university resources, by award year, n = 820   72

8.3   Percent of Phase II NASA-funded projects that
        involved university resources, by award year, n = 306   73

8.4   Percent of Phase II DOE-funded projects that
        involved university resources, by award year, n = 315   73

8.5   Percent of Phase II NSF-funded projects that
        involved university resources, by award year, n = 316   74

12.1 A framework to identify unanticipated consequences 102

12.2 Trends in patents received by DOD-funded firms
based on Phase II research, by commercialization
status, n = 1,137 105

12.3 Trends in patents received by NIH-funded firms
based on Phase II research, by commercialization
status, n = 661 106

12.4 Trends in patents received by NASA-funded firms
based on Phase II research, by commercialization
status, n = 235 107

12.5 Trends in patents received by DOE-funded firms
based on Phase II research, by commercialization
status, n = 249 108

12.6 Trends in patents received by NSF-funded firms
based on Phase II research, by commercialization
status, n = 265 109

Tables

2.1 Legislative history of the purpose statement of the SBIR program 12

3.1 Definition of terms relevant to U.S. technology policy 24

3.2 Characteristics of other elements of U.S. technology policy 25

4.1 Number of Phase II projects in the combined NRC database, by funding agency 32

4.2 Empirical studies of SBIR projects using NRC databases, by publication year 33

6.1 Definitions of internal elements of productive capital supported by data from the NRC databases 47

6.2 Descriptive statistics on internal elements of productive capital, by funding agency 48

6.3 Correlations among DOD-funded firms' internal elements of productive capital, n = 1,463 49

6.4 Correlations among NIH-funded firms' internal elements of productive capital, n = 820 49

6.5 Correlations among NASA-funded firms' internal elements of productive capital, n = 306 50

6.6 Correlations among DOE-funded firms' internal elements of productive capital, n = 315 50

6.7 Correlations among NSF-funded firms' internal elements of productive capital, n = 316 51

7.1 Percent of Phase II projects that failed, by funding agency 55

7.2 Correlations among DOD-funded firms' internal
elements of productive capital and project failure,
n = 1,463   59

7.3 Correlations among NIH-funded firms' internal
elements of productive capital and project failure, n = 820   60

7.4 Correlations among NASA-funded firms' internal
elements of productive capital and project failure, n = 306   60

7.5 Correlations among DOE-funded firms' internal
elements of productive capital and project failure, n = 315   61

7.6 Correlations among NSF-funded firms' internal
elements of productive capital and project failure, n = 316   61

7.7 Percent of Phase II projects that failed, by primary
reason for being discontinued and by funding agency   64

8.1 Percent of Phase II projects that involved
university resources, by funding agency   71

8.2 Correlations among DOD's university productive
capital and project failure, n = 1,463   75

8.3 Correlations among NIH's university productive
capital and project failure, n = 820   76

8.4 Correlations among NASA's university productive
capital and project failure, n = 306   76

8.5 Correlations among DOE's university productive
capital and project failure, n = 315   76

8.6 Correlations among NSF's university productive
capital and project failure, n = 316   77

9.1 Descriptive statistics and correlations among
patents received per year and scientific
publications per year from DOD projects and
productive capital, n = 1,137   82

9.2   Descriptive statistics and correlations among
        patents received per year and scientific
        publications per year from NIH projects and
        productive capital, n = 661   83

9.3   Descriptive statistics and correlations among
        patents received per year and scientific
        publications per year from NASA projects and
        productive capital, n = 235   83

9.4   Descriptive statistics and correlations among
        patents received per year and scientific
        publications per year from DOE projects and
        productive capital, n = 249   84

9.5   Descriptive statistics and correlations among
        patents received per year and scientific
        publications per year from NSF projects and
        productive capital, n = 265   84

10.1   Descriptive statistics on the commercialization
        from a Phase II project, by funding agency   89

10.2   Relationship between stakeholders and the
        commercialization of DOD-funded Phase II
        projects, n = 1,137   90

10.3   Relationship between stakeholders and the
        commercialization of NIH-funded Phase II
        projects, n = 661   91

10.4   Relationship between stakeholders and the
        commercialization of NASA-funded Phase II
        projects, n = 235   91

10.5   Relationship between stakeholders and the
        commercialization of DOE-funded Phase II
        projects, n = 249   92

10.6   Relationship between stakeholders and the
        commercialization of NSF-funded Phase II
        projects, n = 265   92

11.1   Percent of sales of products, processes, and
       services from DOD-funded Phase II technologies,
       by sector, n = 625   95

11.2   Percent of sales of products, processes, and
       services from NIH-funded Phase II technologies,
       by sector, n = 29   96

11.3   Percent of sales of products, processes, and
       services from NASA-funded Phase II technologies,
       by sector, n = 140   96

11.4   Percent of sales of products, processes, and
       services from DOE-funded Phase II technologies,
       by sector, n = 7   96

11.5   Percent of sales of products, processes, and
       services from NSF-funded Phase II technologies,
       by sector, n = 157   97

13.1   Percent of Phase II projects that would have been
       undertaken in the absence of SBIR funding   113

13.2   Characteristics of firms that would have (n = 127) /
       would not have (n = 1,010) undertaken the Phase II
       project in the absence of DOD SBIR funding   114

13.3   Characteristics of firms that would have (n = 85) /
       would not have (n = 576) undertaken the Phase II
       project in the absence of NIH SBIR funding   115

13.4   Characteristics of firms that would have (n = 29) /
       would not have (n = 206) undertaken the Phase II
       project in the absence of NASA SBIR funding   115

13.5   Characteristics of firms that would have (n = 25) /
       would not have (n = 224) undertaken the Phase II
       project in the absence of DOE SBIR funding   116

13.6   Characteristics of firms that would have (n = 36) /
       would not have (n = 229) undertaken the Phase II
       project in the absence of NSF SBIR funding   116

13.7   Implications of undertaking the Phase II project in
        the absence of SBIR funding   118

13.8   Commercialization success of undertaking and not
        undertaking the Phase II project in the absence of
        SBIR funding   119

About the authors

Albert N. Link is the Virginia Batte Phillips Distinguished Professor of Economics at the University of North Carolina at Greensboro (UNCG). He received a B.S. degree in Mathematics from the University of Richmond (Phi Beta Kappa) and a Ph.D. degree in Economics from Tulane University. After receiving his Ph.D., he joined the economics faculty at Auburn University, was later Scholar-in-Residence at Syracuse University, and then he joined the economics faculty at UNCG in 1982. In 2019, Link was awarded the title and honorary position of Visiting Professor at Northumbria University, U.K.

Professor Link's research focuses on technology and innovation policy, the economics of R&D, and policy/program evaluation. He is currently the Editor-in-Chief of the *Journal of Technology Transfer*. He is also co-editor of *Foundations and Trends in Entrepreneurship* and founder/editor of *Annals of Science and Technology Policy*.

Among his more than 70 authored and edited books, some of the more recent ones are: *The Economics and Science of Measurement: A Study of Metrology* (Routledge, 2022), *Technology and Innovation Policy: An International Perspective* (Edward Elgar, 2021), *Invention, Innovation and U.S. Federal Laboratories* (Edward Elgar, 2020), *Technology Transfer and U.S. Public Sector Innovation* (Edward Elgar, 2020), *Collaborative Research in the United States: Policies and Institutions for Cooperation among Firms* (Routledge, 2020), *Sources of Knowledge and Entrepreneurial Behavior* (University of Toronto Press, 2019), *Handbook for University Technology Transfer* (University of Chicago Press, 2015), *Public Sector Entrepreneurship: U.S. Technology and Innovation Policy* (Oxford University Press, 2015), *Bending the Arc of Innovation: Public Support of R&D in Small, Entrepreneurial Firms* (Palgrave Macmillan, 2013), *Valuing an Entrepreneurial Enterprise* (Oxford University Press, 2012), *Public Goods, Public Gains: Calculating the Social Benefits of Public R&D* (Oxford University Press, 2011), *Employment Growth from Public Support of Innovation in Small Firms* (W.E. Upjohn Institute for Employment Research, 2011), and *Government as Entrepreneur* (Oxford University Press, 2009).

Professor Link's other research endeavors consist of more than 250 peer-reviewed journal articles and book chapters, as well as numerous U.S. government reports. His scholarship has appeared in such academic journals as the *American Economic Review*, the *Journal of Political Economy*, the *Review of Economics and Statistics, Economica, Research Policy, Economics of Innovation and New Technology*, the *European Economic Review, Small Business Economics, ISSUES in Science and Technology, Science and Public Policy, Scientometrics*, and the *Journal of Technology Transfer*.

Professor Link's public service includes being a member of the National Research Council's research team that conducted the 2010 evaluation of the U.S. Small Business Innovation Research (SBIR) program. Based on that assignment, he testified before the U.S. Congress in April 2011 on the economic benefits associated with the SBIR program. Link also served from 2007 to 2012 as a U.S. Representative to the United Nations (in Geneva, Switzerland) in the capacity of co-vice chairperson of the Team of Specialists on Innovation and Competitiveness Policies Initiative for the Economic Commission for Europe. In October 2018, Link delivered the European Commission Distinguished Scholar Lecture at the European Commission's Joint Research Centre (in Seville, Spain). Currently, Link is an active member of the National Institute of Standards and Technology funded research team studying the economic impacts of investments in U.S. neutron research sources and facilities.

Martijn van Hasselt is Associate Professor of Economics at the University of North Carolina at Greensboro (UNCG). He received a B.A. (Drs.) degree in Econometrics from the University of Groningen, The Netherlands, and a Ph.D. in Economics from Brown University. After completing his Ph.D., he worked as Assistant Professor of Economics at the University of Western Ontario, and as a Research Economist with RTI International. He has been on the faculty at UNCG since 2014.

Professor van Hasselt has published in the areas of applied econometrics, technology and innovation, and health services research. His research has appeared in a wide range of academic journals, including *Journal of Econometrics, Journal of Applied Econometrics, Economics Letters, Small Business Economics*, and *Economics of Innovation and New Technology*. His current and more recent research focuses on Bayesian modeling of misclassification and endogeneity, with applications in policy evaluation and epidemiology.

Acknowledgments

We thank the National Research Council of the U.S. National Academies of Sciences, Engineering, and Medicine for making available the data analyzed herein. Without their support of our many studies of the SBIR program, this book could not have been written.

Our sincere thanks also go to all of the individuals at Edward Elgar Publishing who were involved throughout the conceptualization stage, the review process, and the publication process. Our sincere thanks too for the comments and suggestions offered by the anonymous reviewers.

Finally, the support of our families as this project progressed has been invaluable.

Abbreviations

D	Development
DHS	Department of Homeland Security
DOC	Department of Commerce
DOD	Department of Defense
DOE	Department of Energy
DOT	Department of Transportation
EC	Experiential capital
ED	Department of Education
EPA	Environmental Protection Agency
GAO	Government Accountability Office
HC	Human capital
HHS	Health and Human Services
IO	Innovative output
IPO	Initial public offering
Kn	Knowledge
NASA	National Aeronautics and Space Administration
NIH	National Institutes of Health
NRC	National Research Council
NSF	National Science Foundation
OECD	Organisation for Economic Co-operation and Development
P.L.	Public Law
r	Rate of return
R	Research
R&D	Research and Development
R&E	Research and Experimentation
SBA	Small Business Administration

SBIR	Small Business Innovation Research
SC	Social capital
STTR	Small Business Technology Transfer
TC	Technical capital
UC	University capital
U.S.	United States
USDA	U.S. Department of Agriculture

PART I

SBIR program: institutional perspectives

1. Introduction to *Small Firms and U.S. Technology Policy*

Many scholars have attributed the growing interest in small firms to a research report published by the Massachusetts Institute of Technology's Neighborhood and Regional Change program in 1979. Birch (1979) concluded in that report, and elsewhere (Birch, 1981), that three-fifths of the net new jobs in the United States between 1969 and 1976 were attributable to small firms with 20 or fewer employees (1979, p. 29):

> On the average about 60 percent of all jobs in the U.S. are generated by independent, small entrepreneurs. Large firms (those with over 500 employees) generate less than 15 percent of all new jobs.

And Birch (1981, p. 8) was emphatic that "large firms are no longer the major providers of new jobs for Americans."

Coincidentally, Birch's report was published during a period of productivity decline in the United States, as discussed in Chapter 2, and just prior to the passage of the Small Business Innovation Development Act of 1982, which legislated the establishment of the Small Business Innovation Research (SBIR) program, as discussed in Chapter 3.

This book is about the SBIR program. While much has been written about the SBIR program from both an institutional perspective and a policy perspective, there remains in our view a conspicuous void of general information about the firms and research projects that are funded through the SBIR program. We provide in this book a multi-dimensional descriptive picture of such firms and their projects to help the reader appreciate what we refer to as the social benefits of the program.

The descriptive nature of many of the chapters that follow needs an explanation. As we discuss in Chapter 4, Congress' interest in small firms, especially the role of small firms in the innovation process, seemed to have peaked in the early 2000s. Congress relied on the National Research Council (NRC) within the U.S. National Academies of Sciences, Engineering, and Medicine (hereafter, the National Academies)[1] to study the SBIR program in detail. The NRC's initial study culminated in the passage of the Small Business Innovation Research Program Reauthorization Act of 2000 (Public Law 106–554).[2] Subsequent studies have been authorized by Congress.

A number of researchers have accessed the project-level data assembled from the NRC studies, and their findings are informative and often enlightening.[3] However, as is often the case in many statistically and econometrically oriented studies, subtleties and nuances about firms funded through the SBIR program, and about their funded research projects in particular, are not presented.[4] Herein, we have attempted to fill that void by exploring new statistical relationships in a manner accessible to the reader. And, in addition to filling that void, the descriptive information in this book about the SBIR program and about program and project relationships not previously considered in the literature might serve as a springboard for future in-depth analyses about the program and its impact on economic and social matters.

Figure 1.1 illustrates how the remaining chapters of this book are organized. The point of origin of this figure is the SBIR program, and the chapters in the book fall within three major sections. Part I of the book, which includes this chapter, discusses the SBIR program from an institutional perspective, Part II of the book discusses the program from an empirical perspective, and Part III of the book discusses the SBIR program from a policy perspective. These sections are labeled in Figure 1.1 and below each are several of the topics discussed.

Figure 1.1   The organization of the book

The legislative history of the SBIR program is discussed in Chapter 2.

In Chapter 3, the SBIR program is placed within the broader context of being an element within the technology policy of the United States. The subtitle of this book is Social Benefits of the U.S. Small Business Innovation Research Program. To date, few of the previous studies about the SBIR program have focused on associated social benefits. Our first characterization of social benefits associated with the SBIR program is presented, and we refer to these benefits by the term *policy transfers*. To date, the concept of an SBIR program, and in some instances the structure of the U.S. SBIR program, have been adopted by nine other countries.

Part II of the book discusses the SBIR program from an empirical perspective. Dimensions from the NRC studies mentioned above have been codified into databases, and those databases are described in detail. The published literature that is, for the most part, based on an empirical analysis of the SBIR program using these databases is summarized in Chapter 4.

From the NRC databases, the so-called demographics of the SBIR program are discussed in Chapter 5, with an emphasis on the gender and minority ownership of firms that received the Phase II awards described in Chapter 2.[5]

In Chapter 6, three constructed variables are introduced, and we refer to these internal (i.e., internal to the researching firm) measures or endowments by the term *productive capital*. These metrics are correlated with other performance metrics in later chapters.

Not all SBIR-funded research is successful; some projects are discontinued before being completed. In Chapter 7, we refer to such projects as failed projects, and the likelihood of a project failing is shown empirically to be inversely related to the presence of productive capital within the firms. That is, the greater a firm's endowments of internal productive capital, the less likely the firms' Phase II research project will fail. This is an important chapter because Phase II projects that failed are removed from the analysis in the remaining chapters.

Chapter 8 identifies yet another productive capital variable—an external one—and that metric is associated with whether or not the firm incorporated university resources in its SBIR-funded Phase II research project. When such external resources are used, they enhance the likelihood that the research project will not fail; that is, that the Phase II research project will be completed and that the newly developed technology will be commercialized.

Our second and third characterizations of social benefits associated with the SBIR program are related to the technology transfers that occur from Phase II research projects. These transfers include both knowledge transfers of the underlying characteristics of the newly developed technology as well as the commercialization of the technology itself, which bring about an increase in consumer surplus. In Chapter 9, we focus on new knowledge being revealed and transferred through patented technology-based information as well as through scientific publications. The volume of such knowledge-based outputs, which we refer to by the term *knowledge transfers* from SBIR projects, is shown to be positively related, on an agency-by-agency basis, to productive capital measures.

In Chapter 10, our attention turns from knowledge transfers to the per se transfer of the newly developed and commercialized technologies from the Phase II research projects. Thus, our fourth characterization of social benefits associated with the SBIR program relates to newly developed technologies entering the marketplace, and we refer to these benefits by the term *technology transfers*. As discussed in Chapter 2, a purpose of the SBIR program is to "increase private-sector commercialisation of innovations [i.e., new technologies] derived from federal research and development funding [i.e., from Phase II research awards]." Our metric for the transfer of the newly developed technology is dichotomous and reflects whether or not the Phase II technology was commercialized. The correlate that we focus on is the presence of a private-sector financial stakeholder (which we referred to in Chapter 6 as social capital) in the firm's technology under development. The presence of a non-SBIR financial stakeholder is positively associated with the commercialization of the Phase II technology.[6]

Given the commercialization of Phase II technologies as a hallmark policy-related accomplishment, a logical question to ask is to whom these new technologies are being sold. We show descriptively in Chapter 11 that most of the newly developed technologies are being sold to (i.e., are being transferred to) organizations in the private sector of the U.S. economy.

Part III of the book discusses the SBIR program from a policy perspective. The descriptive analysis in the previous several chapters has been presented in a positive or objective manner. In Chapter 12, we introduce a normative dimension to our study of the SBIR program by addressing any unanticipated consequences associated with the program. The unanticipated consequences that we address are the patenting of publicly funded research technologies by private-sector firms. While we do not offer a normative or judgmental conclusion about the social benefits of this practice, since we have in Chapter 9 referred to patents as a form of knowledge transfer, we do nevertheless raise the issue of publicly funded technologies becoming open-source technologies for policy makers to consider as part of their continuing due diligence to improve the SBIR program and the social benefits it produces.

Chapter 13 continues with our normative focus on the SBIR program through what we call a counterfactual analysis of funded Phase II research projects. In particular, we rely on NRC survey data to identify those Phase II projects that are self-reported to have been undertaken in the absence of SBIR funding. We show descriptively that such projects would have commercialized on their own, but commercialization would have occurred more slowly in the absence of SBIR funding, where "slowly" means about a year or more. Thus, one might objectively conclude that in these cases the role of SBIR funding was primarily to accelerate the process of bringing new technology to market.

Our concluding remarks are in Chapter 14, and there we summarize the social benefits associated with the SBIR program that we identified and discussed in the previous chapters—the policy transfer of the U.S. SBIR program to other countries, the knowledge transfer of technical knowledge through patents and scientific publications, and the technology transfer of commercialized research outputs—and we suggest a research roadmap for future studies.

NOTES

1. "The National Research Council was organized by the National Academy of Sciences in 1916 to associate the broad community of science and technology with the Academy's purposes of further knowledge and advising the federal government. The Council has become the principal operating agency of both the National Academy of Sciences and the National Academy of Engineering in providing services to the government, the public, and the scientific and engineering communities." See https://history.aip.org/phn/21511003.html. Accessed January 26, 2022.
2. It should not be overlooked either that other countries have adopted programs similar to the U.S. SBIR program, as discussed in Chapter 3 and again in Chapter 14.
3. Many of these papers have been collected and republished in Link (2013).
4. We are not suggesting that such are purposeful decisions on the part of researchers. Rather, journal editors often impose space constraints on authors, and empirical findings are frequently preferred over anecdotal information.

5. A Phase II SBIR-funded research project continues the funded Phase I research which established the technical merit, feasibility, and commercial potential of a proposed research project. During Phase II, the funded research is expected to result in a new commercializable technology.

6. A review of our summary in Chapter 4 of the extant SBIR empirical literature shows that our discussion of financial stakeholders is new to the literature.

2. Legislative history of the SBIR program

SETTING THE STAGE

The United States, like many industrial countries, experienced a productivity slowdown in the early 1970s and then again in the late 1970s and early 1980s. These periods of productivity slowdown are illustrated in Figures 2.1 and 2.2. Figure 2.1 shows the level of the multifactor productivity index for the U.S. private business sector over the years 1948 through 2019, and Figure 2.2 shows the annual percentage change in that index over the same time period. The hashed columns in the figures emphasize the periods of productivity slowdown.[1]

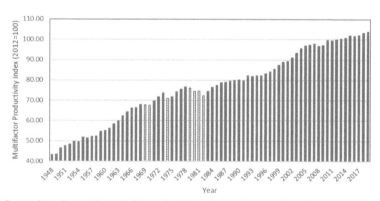

Source: https://www.bls.gov/mfp/mprdload.htm. Accessed January 25, 2022.

Figure 2.1   *Multifactor productivity index for the U.S. private business sector (2012 = 100)*

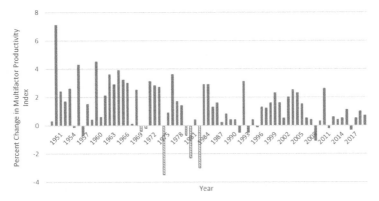

Source: https://www.bls.gov/mfp/mprdload.htm. Accessed January 25, 2022.

Figure 2.2　　*Year-to-year percentage change in multifactor productivity index for the U.S. private business sector*

In response to the productivity slowdown, policy makers as well as academic scholars identified private-sector investments in research and development (R&D) as an important target variable to use to initiate a correction in the economy; in fact, industrial investments in R&D had declined in the few years prior to the initial period of the productivity slowdown (Bozeman and Link, 2015). As we discuss in the following chapter, several R&D-related policies were legislated in the early 1980s to increase or leverage private-sector R&D. One such policy that was intended to benefit the amount of R&D spending in small firms was the Small Business Innovation Development Act of 1982 (Public Law 97–219), which has come to be referred to simply as the 1982 Act.

The contribution of small firms to economic growth was well known in policy circles long before the productivity slowdown (and even before the findings of Birch from Chapter 1) and prior to the passage of the 1982 Act. For example, Bush (1945, p. 21) emphasized the importance of supporting access to R&D in small firms in his *Science—the Endless Frontier*:[2]

> The benefits of basic research do not reach all industries equally or at the same speed. Some small enterprises never receive any of the benefits. It has been suggested that the benefits might be better utilized if "research clinics" for such enterprises were to be established. Businessmen would thus be able

to make more use of research than they now do. This proposal is certainly worthy of further study.

Through the Small Business Act of 1953 (Public Law 85–536), Congress had singled out small business for special treatment—not necessarily for R&D treatment alone—to "strengthen the overall economy":

> It is the declared policy of the Congress that the Government should aid, counsel, assist, and protect, insofar as is possible, the interests of small-business concerns in order to preserve free competitive enterprise ... and to maintain and strengthen the overall economy of the Nation.

Finally, President Jimmy Carter, through his 1979 Domestic Policy Review, emphasized in a message before Congress prior to any legislated productivity slowdown-related actions the innovation-related importance of small firm technologies to the economy (Joint Hearings, 1979, p. 65):

> Small innovative firms have historically played an important role in bringing new technologies into the marketplace. They are also an important source of new jobs. Although many of the initiatives in this Message will encourage such companies, I will also implement several initiatives focused particularly on small firms.
>   First, I propose the enhancement of $10 million of the Small Business Innovation Program of the National Science Foundation. This program supports creative, high-risk, potentially high-reward research performed by small business. Further, the National Science Foundation will assist other agencies in implementing similar programs, with total Federal support eventually reaching $150 million per year.

ENABLING LEGISLATION

The preamble to the Small Business Innovation Development Act of 1982 (Public Law 97–219) reads:

> The Congress finds that:
> (1) technological innovation creates jobs, increases productivity, competition, and economic growth, and is a valuable counterforce to inflation and the United States balance-of-payments deficit;
> (2) while small business is the principal source of significant innovations in the Nation, the vast majority of federally funded research and devel-

opment is conducted by large businesses, universities, and Government laboratories; and

(3) small businesses are among the most cost-effective performers of research and development and are particularly capable of developing research and development results into new products.

And, the purposes of the 1982 Act, and by default the initial purposes of the SBIR program, are also stated to be:

(1) to stimulate technological innovation;
(2) to use small business to meet Federal research and development needs;
(3) to foster and encourage participation by *minority and disadvantaged* [emphasis added] persons in technological innovation; and
(4) to increase private sector commercialization innovations derived from Federal research and development.

To fund these purposes, the Act stated that:[3,4]

Each Federal agency which has an extramural budget for research or research and development in excess of $100,000,000 for fiscal year 1982, or any fiscal year thereafter, shall expend not less than 0.2 per centum of its extramural budget in fiscal year 1983 or in such subsequent fiscal year as the agency has such budget, not less than 0.6 per centum of such budget in the second fiscal year thereafter; not less than 1 per centum of such budget in the third fiscal year thereafter, and not less than 1.25 per centum of such budget in all subsequent fiscal years with small business concerns specifically in connection with a small business innovation research program which meets the requirements of the Small Business Innovation Development Act of 1982 and regulations issued thereunder.

The 1982 Act was not a permanent legislation.[5] Accordingly, the SBIR program has been reauthorized by Congress a number of times. Through these reauthorizations, the *purpose statement* has been modified, and the purpose statement was later expanded into a statement of *program goals*, as discussed below.[6] See Table 2.1 for changes in the purpose statements.

Table 2.1   *Legislative history of the purpose statement of the SBIR program*

Legislation	Public Law (P.L.)	Time Period	Purpose Statement
Small Business Innovation Act of 1982	P.L. 97–219	1982–86	"… the purposes of the Act are— (1) to stimulate technological innovation; (2) to use small business to meet Federal research and development needs; (3) to foster and encourage participation by *minority* and *disadvantaged persons* in technological innovation; and (4) to increase private sector commercialization innovations derived from Federal research and development."
Department of Defense Appropriation Act of 1986	P.L. 99–443	1986–92	unchanged
Small Business Research and Development Enactment Act of 1992	P.L. 102–564	1992–2000	"The purposes of this title are— (1) to expand and improve the small business innovation research program; (2) to emphasize the program's goal of increasing private sector commercialization of technology developed through Federal research and development; (3) to increase small business participation in Federal research and development; and (4) to improve the Federal Government's dissemination of information concerning the small business innovation research program, particularly with regard to program participation by *women-owned* small business concerns and by *socially and economically disadvantaged* small business concerns."
Small Business Reauthorization Act of 2000	P.L. 106–554	2000–08	unchanged
Temporary Extensions	–	2008–12	unchanged
National Defense Authorization Act of 2012	P.L. 112–81	2012–17	unchanged
National Defense Authorization Act of 2017	P.L. 114–328	2017–22	unchanged

Note: Our emphasis of terms and phrases is in italics.

Through the reauthorization legislations of the SBIR program, the mandated set-aside amount has also been changed; the upper-limit amounts for Phase I and Phase II awards (defined below) have been increased.[7] In 2016, the SBIR program was extended through September 30, 2022, by the 2017 National Defense Authorization Act (Public Law 114–328). Currently (as of 2022) the set-aside amount is 3.2 percent of a participating agency's extramural research budget.

With funding mandated by the 1982 Act, the SBIR program was soon organizationally structured with three phases (Phase I, Phase II, and Phase III), two of which are funded through the program (Phase I and Phase II awards).[8]

Phase I competitive awards support research to establish the technical merit, feasibility, and commercial potential of the proposed research and to determine the quality of performance of the small business awardee's organization prior to providing further federal support in Phase II. Currently, Phase I awards generally range from $50,000 to $250,000, and they last for six months.[9] Phase II competitive awards fund the continuation of the Phase I research, and funding is based on the results achieved in Phase I and the scientific and technical merit and commercial potential of the project proposed in competitive Phase II. Currently, Phase II awards are generally capped at $750,000, and their duration is for two years. During Phase III, small businesses are expected to pursue third-party funding to fund the commercialization of the technology resulting from the Phase I and Phase II research. The SBIR program does not fund its currently funded technology-related R&D in Phase III.

There are a number of exceptions to the upper-limit amounts on both Phase I and Phase II project awards. As referenced in Gallo (2021, p. 5), the 2019 *Small Business Innovation Research (SBIR) Program Policy Directive*:[10]

> ... provides agencies with the authority to issue an award that exceeds this amount by as much as 50% ... [A]gencies may request a waiver from the SBA [Small Business Administration] to exceed the Phase II award [and the Phase I award] guideline by more than 50% for a specific topic. In general, the period of performance for Phase II awards is not to exceed two years, though agencies may allow for a longer performance period for a particular project. Agencies may make a sequential Phase II award to continue the work of an initial Phase II award. The amount of a sequential Phase II award is subject to the same Phase II award guideline and agencies' authority to exceed the guideline by up to 50%. Thus, agencies may award up to $3 million, adjusted for inflation, in Phase II awards for a particular project to a single recipient

at the agency's discretion, and potentially more if the agency requests and receives a waiver from the SBA.

To be eligible for an SBIR award, the small business must be: independently owned and operated; other than the dominant firm in the field in which it is proposing to carry out SBIR projects; organized and operated for profit; the employer of 500 or fewer employees, including employees of subsidiaries and affiliates; the primary source of employment for the project's principal investigator at the time of award and during the period when the research is conducted; and at least 51 percent owned by U.S. citizens or lawfully admitted permanent resident aliens. However, there are also agency exceptions where the number of employees can be greater than 500.

The current mission statement of the SBIR program is to:[11]

... support scientific excellence and technological innovation through the investment of Federal research funds in critical American priorities to build a strong national economy.

And the current goals of the program (no longer referred to by the phrase *purpose of the program*) is to:[12]

(1) Stimulate technological innovation.
(2) Meet Federal research and development needs.
(3) Foster and encourage participation in innovation and entrepreneurship by *women* and *socially or economically disadvantaged* [emphasis added] persons.[13]
(4) Increase private-sector commercialization of innovations derived from Federal research and development funding.

Goal (3) just above is different from purpose statement (3) contained in the 1982 Act (see Table 2.1). Although the words *woman* or *women* are not mentioned in the Small Business Act of 1953, gender has become an emphasized part of the amendments to that Act. Currently, the Small Business Act of 1953, as amended through Public Law 117–6 on March 30, 2021, states:

(1)... Congress finds that—
(A) women owned business has become a major contributor to the American economy by providing goods and services, revenues, and jobs;
(B) over the past two decades there have been substantial gains in the

social and economic status of women as they have sought economic equality and independence;

(C) despite such progress, women, as a group, are subjected to discrimination in entrepreneurial endeavors due to their gender;

(D) such discrimination takes many overt and subtle forms adversely impacting the ability to raise or secure capital, to acquire managerial talents, and to capture market opportunities;

(E) it is in the national interest to expeditiously remove discriminatory barriers to the creation and development of small business concerns owned and controlled by women;

(F) the removal of such barriers is essential to provide a fair opportunity for full participation in the free enterprise system by women and to further increase the economic vitality of the Nation;

(G) increased numbers of small business concerns owned and controlled by women will directly benefit the United States Government by expanding the potential number of suppliers of goods and services to the Government; and

(H) programs and activities designed to assist small business concerns owned and controlled by women must be implemented in such a way as to remove such discriminatory barriers while not adversely affecting the rights of socially and economically disadvantaged individuals.

(2) It is, therefore, the purpose of those programs and activities conducted under the authority of this Act that assist women entrepreneurs to—

(A) vigorously promote the legitimate interests of small business concerns owned and controlled by women;

(B) remove, insofar as possible, the discriminatory barriers that are encountered by women in accessing capital and other factors of production; and

(C) require that the Government engage in a systematic and sustained effort to identify, define and analyze those discriminatory barriers facing women and that such effort directly involve the participation of women business owners in the public/private sector partnership.

Eleven public-sector agencies and organizations currently participate in the SBIR program. They are (alphabetically): the Departments of Agriculture (USDA), Commerce (DOC), Defense (DOD), Education (ED), Energy (DOE), Health and Human Services (HHS), Homeland Security (DHS), Transportation (DOT), and the Environmental Protection Agency (EPA), the National Aeronautics and Space Administration (NASA), and the National Science Foundation (NSF).

ECONOMICS OF THE SBIR PROGRAM

An economic argument for public-sector support of private-sector R&D in general is that private-sector firms underinvest in R&D (Arrow, 1962).[14] Of particular concern to the public sector (i.e., to public-sector policy makers and hence to society) is the failure of firms to invest in the socially optimal level of R&D because of investment barriers that bring about what is commonly known by the term *market failure*.

Link and Scott (2012b, 2013) suggest that risk and uncertainty may arise in the R&D production process (i.e., technical risk/uncertainty) or in the process of participating in factor markets (i.e., market risk/uncertainty). Technical risk and uncertainty are attributable to the nature of the R&D process and to the fact that the outcomes of the R&D process are uncertain.[15] Likewise, market risk and uncertainty are attributable to the vicissitudes of participating in the marketplace. An uncertain demand for the firm's output, an uncertain supply for the firm's inputs, an uncertain ability of the firm to appropriate returns from its R&D investments, an uncertain market structure, and so on, are all factors that bring about R&D uncertainty.

Both risk and uncertainty will result in performance difficulties for the firm. In the case of risk (quantifiable random variability), while insurance markets may, if they even exist, allow the risk to be shifted to others, that shifting comes at a price that reduces the return that the firm can expect to receive.[16] It follows that this reduction may result in the firm's expected return falling below its minimum rate of return, that is, its hurdle rate, required by the firm to go forward with the project. Even more problematic is the case of when markets are incomplete and thus associated with high levels of risk. In that case, the risk is sufficiently high that there is no available insurance market. Of course, the firm can decide to internalize the risk; however, in that case, particularly if the firm is risk averse, the effect can be much the same with that risk lowering the expected return below the firm's hurdle rate and thus resulting in the firm choosing not to engage in the innovation project.

In the case of uncertainty (unquantifiable random variability), the result is even more troubling. Because of the lack of knowledge about possible outcomes and their probabilities, no insurance market will exist. In other words, there will be an incomplete markets problem. As a result, the firm will have to bear the effect of the uncertainty. Depending on the firm's subjective estimation of the likelihood of success, this uncertainty may result in the firm deciding not to engage in the innovation process.

In this context, the SBIR program represents a way of correcting for these market failures by helping the firm over its hurdle rate so that it invests in a specific R&D project that it would not have invested in without the award. Link and Scott (2010, 2012b, 2013) explain why this is the case with reference to Figure 2.3. The figure is framed in terms of risk and expected rates of return, hence the use of probability distributions and probabilities.[17]

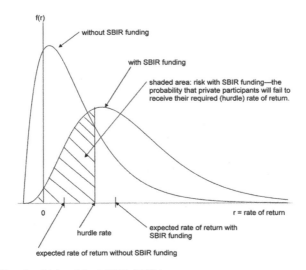

Source: Based on Link and Scott (2010, 2012b).

Figure 2.3   *Risk and expected rate of return for a private-sector firm as a result of SBIR funding*

Figure 2.3 is constructed to represent the reduction of the firm's technical and market risk, referred to in the figure collectively, for simplicity, as risk, from receipt of an SBIR award. Consider first the firm's circumstances in the absence of SBIR funding. In that case, the probability distribution of the rate of return for the private firm's project will be the distribution on the left in the figure. As drawn, the private hurdle rate is to the right of the expected rate of return without SBIR funding, meaning that the private firm will not undertake this research because the firm will not receive its required rate of return.[18] The risk of the project for the firm is the probability of "failure," namely, the probability that the rate of return is less than the firm's hurdle rate. Thus, the risk of the project is equal to the area under the distribution labeled "without-SBIR funding" that is to the left of the private hurdle rate.[19]

Consider the effect of an SBIR award. For a given SBIR award, the expected rate of return associated with any particular outcome of the underlying stochastic process will be higher than without the award. It follows then that an SBIR award will result in a rightward shift in the distribution of the rate of return for the private firm. Because of that rightward shift, the firm's expected private rate of return, as well as the variance in the rate of return, will increase.[20] And with that increase, the downside risk associated with undertaking R&D, illustrated in Figure 2.3 by the shaded region under the distribution, will fall, hence making it more likely, as is illustrated in Figure 2.3, that the firm's expected private rate of return will exceed the firm's private hurdle rate.[21]

Note, however, that SBIR funding will not increase the probability that the research will be completed, assuming the research was in fact undertaken absent SBIR funding. But it will reduce private risk by increasing the expected private rate of return because the expected rate of return will be based on a smaller private outlay. Thus, SBIR funding leverages the private firm's investment as illustrated by a greater expected return and a greater variance in the distribution, as explained above. That in turn increases the overall level of R&D activity for participating firms in the SBIR program. Increasing the overall level of R&D for participating firms is a dimension of the social benefits associated with the SBIR program, a point that we will expand on in future chapters.

In Figure 2.3, the expected rate of return with SBIR funding is greater than the private hurdle rate and from our discussion it was implicitly assumed that the funding project's social rate of return is greater than the social hurdle rate (not shown in Figure 2.3). In Chapter 12, under the umbrella of an unanticipated consequence from the SBIR program, we consider the case where the expected rate of return to the SBIR-funded project is above the private hurdle rate (as shown in Figure 2.3) but the project's social rate of return is below the social hurdle rate.

SUMMARY AND WHAT FOLLOWS

We have provided the institutional context for this book in this chapter. In particular, we emphasized the mission of the SBIR program and its legislated goals. The mission and goals are particularly important when one thinks about the economic benefits associated with the program, and they are important when one thinks about the SBIR program from a policy perspective.

In the following chapter, the SBIR program is discussed in terms of it being an element of U.S. technology policy. The emphasis of Chapter 3 is that many of the elements of U.S. technology policy are, like the SBIR program, designed to leverage private-sector investments in R&D.

NOTES

1. The U.S. Bureau of Labor Statistics calculates this aggregate multifactor productivity index. See https://www.bls.gov/mfp/mprdload.htm. Accessed January 24, 2022. Not shown in Figure 2.1 is the multifactor productivity index for 2020, the first year of COVID-19. At the time of writing this chapter that index was preliminary.
2. See also Link (2022).
3. A prototype of the SBIR program began in 1977 at the National Science Foundation (NSF). Ronald Tibbetts, the so-called Father of the SBIR Program, "was appointed as the Senior Program Officer at NSF in 1972. As an NSF program manager, Tibbetts was known as a task master with well-honed instincts for enabling potentially game-changing projects. He also recognized the importance of small, high-tech firms to the economy and observed the fierce opposition they faced from other recipients when pursuing federal R&D funding ... Senator Edward Kennedy also recognized the vital role that small businesses play in America's growing economy. He spent much of the 1970's tirelessly championing for NSF to support the research of qualified small businesses as the chairman of the National Science Foundation Subcommittee of the Senate Labor and Public Welfare Committee. Kennedy continued to introduce different proposals to increase the percentage of the budget directed toward small businesses. Finally, once NSF recognized the need for ongoing support for small business, the foundation instituted the SBIR Program in 1977. In addition to Senator Kennedy, much of the legislative support for the SBIR Program was directly due to the work of Arthur and Judith Obermayer. As early as 1970, Arthur testified before the U.S. Congress on the challenges small R&D companies faced in dealing with the government. He also lobbied alongside Kennedy for the initial 1974 NSF Authorization Act ..." See https://www.sbir.gov/birth-and-history-of-the-sbir-program. Accessed May 9, 2022.
4. To repeat, a social importance of the U.S. SBIR program might be inferred from the number of other countries that have adopted programs similar to the SBIR program, as will be discussed in Chapter 3 and again in Chapter 14.
5. This section draws directly from Link and Scott (2010, 2012a, 2012b, 2012c, 2013) and Leyden and Link (2015a, 2015b).

6. The National Defense Authorization Act of 2017 reauthorized the SBIR program until 2022, and the *purpose statement* of the program was not changed from the Small Business Research and Development Enactment Act of 1992, the stated *program goals* are now what is emphasized.

7. A history of these legislated changes is detailed in Leyden and Link (2015a, 2015b).

8. See https://www.sbir.gov/about. Accessed January 25, 2022.

9. Phase I research is often referred to as *proof of concept research.* However, researchers and policy makers are likely accustomed to thinking about funded research in terms of the character of use of R&D as defined by NSF. See https://www.nsf.gov/statistics/randdef/ (accessed January 25, 2022): "Basic research is experimental or theoretical work undertaken primarily to acquire new knowledge of the underlying foundations of phenomena and observable facts, without any particular application or use in view. Applied research is original investigation undertaken in order to acquire new knowledge. It is directed primarily towards a specific, practical aim or objective. Development (Experimental development) is systematic work, drawing on knowledge gained from research and practical experience and producing additional knowledge, which is directed to producing new products or processes or to improving existing products or processes." None of these definitions match the scope of research that falls under the rubric of a Phase I project. We thank John Jankowski, Director of the R&D Statistics Program at NSF, for sharing with us his professional view that proof of concept research probably "falls in the end of the applied research spectrum." For reference, the Organisation for Economic Co-operation and Development's (OECD) Frascati Manual does not define proof of concept research. See https://www.oecd-ilibrary.org/science-and-technology/frascati-manual -2015_9789264239012-en. Accessed January 25, 2022.

10. "As of November 2021, agencies may issue a Phase I award (including modifications) up to $275,766 and a Phase II award (including modifications) up to $1,838,436 without seeking SBA approval. Any award above those levels will require a waiver." See https://www.sbir.gov/about. Accessed April 2, 2022. As of January 5, 2022, National Institutes of Health (NIH) may exceed these normal amounts for projects on special topics. See https://grants.nih .gov/grants/guide/notice-files/NOT-OD-22–043.html. Accessed June 22, 2022.

11. See https://www.sbir.gov/about. Accessed January 29, 2022.

12. Ibid.

13. See https://seedfund.nsf.gov/fastlane/definitions/. Accessed January 25, 2022. "A member of any of the following groups: Black Americans, Hispanic Americans, Native Americans, Asian-Pacific Americans, Subcontinent Asian Americans, other groups designated from time to time by the Small Business Administration (SBA) to be socially disadvantaged, and any other individual found to be socially and economically disadvantaged by SBA pursuant to Section 8(a) of the Small Business Act, 15 U.S.C.; 637(a)."

14. This section draws directly from Link and Scott (2010, 2012b, 2013).

15. If the R&D process is uncertain but probabilistically predictable, then it is technical risk; if the R&D process is uncertain so that it is not possible to enumerate all possible outcomes and/or quantify the probabilities of all possible outcomes, then it is technical uncertainty.
16. The remainder of this section draws directly from Link and Scott (2009, 2012b).
17. The logic would be the same in the case of uncertainty; all that is needed is a relabeling of the vertical axis as the subjective likelihood of success and relabeling the probability distributions as subjective likelihood functions.
18. Note that the expected rate of return does not necessarily correspond to the greatest frequency or probability density because the distribution of rates of return is not necessarily symmetric.
19. This notion of downside risk is consistent with Shackle's (1979) analysis of decision making under uncertainty. For those readers who are used to thinking of the variance of the distribution as the measure of risk, this notion of downside risk—which is the probability of a rate of return less than the hurdle rate—might seem unusual. But the nature of the firm's problem is what is important. Variance measures of risk are based on the notion that the firm is concerned with the possibility that outcomes can differ from the expected outcome. But in the case of a firm deciding whether to engage in research, the concern is whether the research project will result in success or failure.
20 As Link and Scott (2010, 2012b) explain, to capture the idea of limited liability for investors, we bound their return below by zero.
21. SBIR funding need not affect the firm's private hurdle rate.

3.   The SBIR program: an element of U.S. technology policy

The SBIR program is one of several elements of U.S. technology policy that is specifically focused on small firms. To place the SBIR program within a policy context, we review in this chapter several elements of U.S. technology policy. As we discussed in Chapter 2, Congress promulgated a number of R&D-focused and technology-focused policies in response to both the productivity slowdown and to President Carter's charges in his 1979 Domestic Policy Review. The vehicle that we use for this purpose is shown in Figure 3.1.

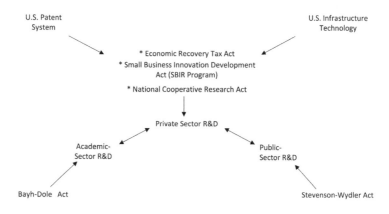

Figure 3.1   Representation of U.S. technology policy

DEFINITIONS

Toward our goal of placing the SBIR program in the context of U.S. technology policy, the terms *technology* and *policy* must be defined, but so must other terms because many related terms (i.e., the terms *technology* and *innovation*) are used interchangeably in common parlance.

Consider the following four definitions:

> *Science*—the search for knowledge; the search is based on observed facts and truths; science begins with known starting conditions and searches for unknown end results.

> *Technology*—the application of knowledge, learned through science, to some practical problem.

> *Innovation*—technology put into use or commercialized.

> *Policy*—a course of action adopted and pursued by an organization, and herein the organization is assumed to be in the public sector, and it is referred to generically as the government.

Our generalization from these definitions is that the term *technology policy* refers to policy to enhance the application of new knowledge that often occurs within an R&D laboratory—a private-sector firm's research laboratory or a public-sector research laboratory. The term *innovation policy* refers to policy to enhance the commercialization of a resulting technology, and that commercialization occurs directly within the marketplace.[1] Thus:

> *Technology policy*—policy to enhance, in a laboratory, the application of knowledge, learned through science, to some known problem.

> *Innovation policy*—policy to enhance the commercialization of a technology within the marketplace.

The above definitions are collectively presented in Table 3.1 for reference purposes. All of the policies emphasized in this chapter have a focus on private-sector R&D.

Table 3.1   *Definition of terms relevant to U.S. technology policy*

Term	Definition
Science	The search for knowledge; the search is based on observed facts and truths; science begins with known starting conditions and searches for unknown end results.
Technology	The application of knowledge, learned through science, to some practical problem.
Innovation	Technology put into use or commercialized.
Policy	A course of action adopted and pursued by an organization, and herein the organization is assumed to be in the public sector, and it is referred to generically as the government.
Technology policy	A policy to enhance, in a laboratory, the application of knowledge, learned through science, to some known problem.
Innovation policy	A policy to enhance the commercialization of a technology within the marketplace.

Source: Based on Cunningham and Link (2021b).

DESCRIPTION OF ELEMENTS OF U.S. TECHNOLOGY POLICY

The four technology policies that are represented in Figure 3.1, in addition to the Small Business Innovation Development Act of 1982, are: the Bayh-Dole Act of 1980, the Stevenson-Wydler Act of 1980, the Economic Recovery Tax Act of 1981, and the National Cooperative Research Act of 1984. Table 3.2 provides a brief summary of these four technology policies with an emphasis on the segment(s) of the economy that each policy is intended to affect.

The point of comparative emphasis from Table 3.2 is that the SBIR program is not the only element of U.S. technology policy that focuses on small private-sector firms, but it is the element that focuses *exclusively* on small private-sector firms.[2] That said, it is possible for a tax-paying small firm to benefit from the research and experimentation (R&E) tax credit,[3] and it is possible for a small firm to be involved in cooperative research with another firm or firms, but the exclusivity of emphasis on small firms is only relevant to the SBIR program.

Table 3.2   *Characteristics of other elements of U.S. technology policy*

Policy	Description	Affected Segments
Bayh-Dole Act of 1980 (formally, the University and Small Business Patent Protection Act of 1980)	Redefined property rights to facilitate the transfer of existing knowledge resulting from public-sector funded research in universities to the private sector	Universities and private-sector firms
Stevenson-Wydler Act of 1980 (formally, the Stevenson-Wydler Technology Innovation Act of 1980)	Called for federal laboratories to promote technology transfer to the private sector for commercial exploitation. Each federal laboratory was required to establish an Office of Research and Technology Application to facilitate the technology transfer.	Federal laboratories (and indirectly private-sector firms)
R&E Tax Credit portion of the Economic Recovery Tax Act of 1981	Provided a marginal tax credit to private-sector firms on qualified R&E expenditures in excess of the average amount spent during the previous three taxable years.	Private-sector research-active firms
National Cooperative Research Act of 1984	Created a registration process under which joint R&D ventures can voluntarily disclose their research intension to the U.S. Department of Justice and thereby gain partial indemnification from antitrust laws and penalties.	Private-sector firms and their research partners

Note: A federal laboratory is any laboratory—a federally funded research and development center, or center that is owned, leased, or otherwise used by a federal agency and funded by the federal government—operated by the government or by a contractor.

SBIR PROGRAM AS AN EXAMPLE OF PUBLIC-SECTOR ENTREPRENEURSHIP

The term *public-sector entrepreneurship* is defined as (Leyden and Link, 2015a, p. 14):

> [P]ublic sector entrepreneurship refers to innovative public policy initiatives that generate greater economic prosperity by transforming a *status quo* economic environment into one that is more conducive to economic units engaging in creative activities in the face of uncertainty.

We suggest that the Small Business Innovation Development Act, and the SBIR program in particular, are examples of public-sector entrepreneurship because the program is an innovative public policy that transforms "a *status quo* economic environment" into one that is conducive to firms "engaging in creative activities in the face of uncertainty."

The SBIR program is innovative in at least two dimensions. First, it is focused exclusively on small firms, and second, it is structured in a way to encourage small firms to engage in R&D projects that they would not otherwise undertake because of the uncertainty of the research process, a point that we revisit in Chapter 13. And as a result, funded firms have an opportunity at commercial success through the development of a new technology.

To emphasize the innovative nature of the SBIR program, a number of other countries have adopted programs similar to the U.S. program. Link and Cunningham (2021) document the global adoption of such programs: South Africa in 1993, Turkey in 1995, Australia in 1996, South Korea in 1998, Japan in 1999, Taiwan in 1999, United Kingdom in 2001, The Netherlands in 2005, and New Zealand in 2012.

The adoption of SBIR-like programs by these nine countries is a social benefit associated with the program. Specifically, this policy transfer facilitates these countries' ability to internalize, for the purpose of economic growth, knowledge transfers and technology transfers from their own small firms. The diffusion of the SBIR program across countries is illustrated in Figure 3.2.

The upper portion of Figure 3.2 shows the nine countries that have adopted SBIR-like programs by year of adoption. The lower portion of the figure illustrates what might be called an arrival model, that is, the timing of the arrival of an SBIR-like program in the country. The cumulative proportion of adopting countries is measured on the vertical axis. Imposed on the jagged line diagram is a third-degree polynomial (dashed) to illustrate a traditional lazy-S diffusion or arrival curve of the imitated policy.

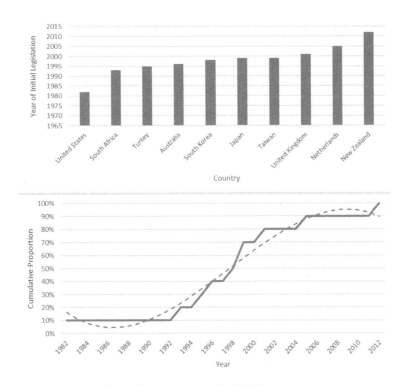

Figure 3.2   The policy transfer of the SBIR program to other countries

SUMMARY AND WHAT FOLLOWS

In this chapter we placed the SBIR program within the context of the broader technology policy of the United States. And, we emphasized that an understanding of the social benefits of the program are appropriately viewed in term of the segments of the economy that are affected by the SBIR program.

In Chapter 4, the databases used in the subsequent chapters to describe aspects of the SBIR program are described not only in terms of their genesis development but also in terms of their coverage of Phase II projects. Also, the extant literature that has previously been based on these data is reviewed in an effort to codify the research questions that have been asked and answered by other researchers.

NOTES

1. These definitions explain the commonly used policy phrase, *from lab to market*.
2. A number of states also have programs to support research in small firms. For example, see Hardin et al. (2020).
3. R&E refers to research and experimentation expenditures. Such expenditures include "expenditures incurred in connection with the taxpayer's trade or business which represent research (R) and development (D) costs in the experimental or laboratory sense." Generally, for a firm, R&E expenditures are less than its R&D expenditures because not all development (D) expenditures are experimental. See https://www.law.cornell.edu/cfr/text/26/1.1742. Accessed January 26, 2022.

PART II

SBIR program: empirical perspectives

4. SBIR databases and empirical studies of the SBIR program

SBIR DATABASES

Much of the empirical academic and policy literature that is related to the SBIR program has been based on data collected by the NRC within the National Academies. As background, the NRC was charged in 2000 with the collection of SBIR program-specific data. The NRC wrote in its report to Congress (NRC, 2008, p. 13):

> As the Small Business Innovation Research (SBIR) program approached its twentieth year of operation, the U.S. Congress asked the National Research Council (NRC) to carry out a "comprehensive study of how the SBIR program has stimulated technological innovation and used small businesses to meet federal research and development needs" and make recommendations on improvements to the program.

More formally, authority for this NRC study, which has become known as the first-round assessment of the SBIR program, was a part of the Small Business Innovation Research Program Reauthorization Act of 2000 (Public Law 106–554):

> Congress finds that—
> (1) the small business innovation research program established under the Small Business Innovation Development Act of 1982, and reauthorized by the Small Business Research and Development Enhancement Act of 1992 … is highly successful in involving small businesses in federally funded research and development;
> (2) the SBIR program made the cost-effective and unique research and development capabilities possessed by the small businesses of the Nation available to Federal agencies and departments;
> (3) the innovative goods and services developed by small businesses that participated in the SBIR program have produced innovations of criti-

cal importance in a wide variety of high-technology fields, including biology, medicine, education, and defense;

(4) the SBIR program is a catalyst in the promotion of research and development, the commercialization of innovative technology, the development of new products and services, and the continued excellence of this Nation's high-technology industries; and

(5) the continuation of the SBIR program will provide expanded opportunities for one of the Nation's vital resources, its small businesses, will foster invention, research, and technology, will create jobs, and will increase this Nation's competitiveness in international markets. The Administration is to collect such information from awardees as is necessary to assess the SBIR program …

Among other activities in response to its Congressional charge, the NRC collected information from Phase II awardees funded by DOD, NIH, NASA, DOE, and NSF over the fiscal year time period 1992 through 2001 on a number of aspects of the awardees and on dimensions of their funded Phase II research projects.[1] The collected information was codified into what we call the 2005 Database, and it was made available for study by researchers who are referenced in Table 4.2.

In Section 5137, Continued Evaluation by the National Academy of Sciences, of the National Defense Authorization Act for Fiscal Year 2012 (Public Law 112–81), Congress again charged the NRC (National Academies, 2016, p. 21):[2]

… to review the Small Business Innovation Research and Small Business Technology Transfer (SBIR/STTR) programs at the Department of Defense, the National Institutes of Health, the National Aeronautics and Space Administration, the Department of Energy, and the National Science Foundation. Building on the outcomes from the first-round study, this second-round [assessment] study is to examine both topics of general policy interest that emerged during the first-round study and topics of specific interest to individual agencies.

As was done for the first-round study, the NRC again collected information from Phase II awardees on a number of aspects of the awardees and on their funded research projects. In 2011, as the first part of the second-round study of the SBIR program, Phase II projects funded by DOD, NASA, and NSF were surveyed; in 2014, Phase II projects funded by NIH and DOE were surveyed. The survey information from these agency studies was codified into what we call, respectively, the 2011 Database and the 2014 Database.

For the construction of the 2005 Database, the 2011 Database, and the 2014 Database, the NRC implemented a process to select a random sample of Phase II awards from the five agencies with the largest SBIR programs: DOD, NIH, NASA, DOE, and NSF.[3] For reference, as of the end of fiscal year 2020, the distribution of all Phase II awards by agency program was (rounded): DOD, 52.1 percent; NIH, 22.6 percent; NASA, 6.5 percent; DOE, 9.1 percent; NSF, 5.3 percent; and all other agencies, 4.3 percent.[4]

The total number of Phase II awards assembled into the 2005 Database, the 2011 Database, and the 2014 Database are presented in Table 4.1.[5] The years covered for each agency differ based on the timing of the 2011 and 2014 surveys. For DOD, the years covered are 1992–2007; for NIH, the years covered are 1992–2010; for NASA, the years covered are 1992–2007; for DOE, the years covered are 1992–2010; and for NSF, the years covered are 1992–2009.

Table 4.1   *Number of Phase II projects in the combined NRC database, by funding agency*

Funding Agency	Number of Phase II Projects
DOD	1,685
NIH	1,068
NASA	360
DOE	401
NSF	571
Total	4,085

An upside of using the merged databases is that they provide a larger sample of Phase II projects to study and to use for descriptive purposes; a downside is that the award amounts for Phase II projects in the 2011 Database and the 2014 Database were not made available to researchers in an effort by the NRC to maintain firm privacy. Many of the studies summarized in Table 4.2 that used project and firm information from the 2005 Database are thus not reproducible with the 2011 Database and the 2014 Database. That said, based on an analysis of the 2005 Database, there is a statistically significant and positive relationship between the number of employees in the firm when the Phase II award was submitted and the amount of the Phase II award.[6] Thus, in the analysis that follows, the number of employees in the firm, which is referred to as a measure of human capital in Chapter 6, is perhaps also a reasonable proxy for the resource bases (a human capital base and an R&D or technical capital base) of the firm.

EMPIRICAL STUDIES

Table 4.2 summarizes the empirical studies related to the SBIR program based on either the 2005 Database, the 2011 Database, or the 2014 Database. Conspicuously absent from the summary of empirical studies in Table 4.2 is a study that combines information from the three databases in an effort to explore in greater statistical detail the social benefits of the SBIR program over time. The analyses in this book are, to the best of our knowledge, the first systematic effort to begin to fill that void. By doing so, we might also be taking a first step on how subsequently collected Phase II project data might be studied.

Table 4.2   *Empirical studies of SBIR projects using NRC databases, by publication year*

Author(s)	Key Findings	NRC Data
Audretsch, Link, and Scott (2002)	The authors calculate that the net economic benefits associated with the DOD SBIR program are positive.	Survey data collected as part of NRC's first-round assessment
Link and Ruhm (2009)	The authors quantify empirically the probability of a NIH-funded Phase II project being commercialized is functionally related to the project's ability to attract Phase III development funds.	2005 Database
Link and Scott (2009)	The authors conclude from their analysis of DOD Phase II projects that the probability of a project resulting in a commercialized innovation is less than 50%. The authors suggest that a prediction market could improve that performance.	2005 Database
Bearse and Link (2010)	The authors find the probability that a funded Phase II project will be completed, and thus meet its technical goals, will increase by 8 percentage points when funding is increased from the $750,000 threshold to $1,000,000.	2005 Database

Author(s)	Key Findings	NRC Data
Link and Scott (2010)	The authors introduce a theoretical construct shown in Figure 2.3, and they quantify empirically a host of covariates associated with the commercialization activity of Phase II projects funded by DOD, NIH, NASA, DOE, and NSF.	2005 Database
Link and Ruhm (2011)	The authors conclude that firms conducting Phase II research funded by DOD, NIH, NASA, DOE, and NSF tradeoff between publishing and patenting their research findings, and that tradeoff is functionally related to the intellectual backgrounds of founders. Firms with founders with academic backgrounds publish relatively more; firms with founders with business backgrounds patent relatively more.	2005 Database
Allen, Layson, and Link (2012)	The authors infer from their analysis of the producer and consumer surplus associated with Phase II awards funded by DOD, NIH, NASA, DOE, and NSF that the benefit-to-cost ratio for the programs is greater than unity.	2005 Database
Audretsch, Leyden, and Link (2012)	The authors find from their analysis of DOE Phase II projects that larger firms and firms with founders with an academic background are more likely to be involved in a research partnership with a university.	2005 Database
Link and Scott (2012a)	Based on a model of employment growth, the authors estimate that employment gains from Phase II DOD, NIH, NASA, DOE, and NSF projects have been numerically large but not statistically significant in most cases.	2005 Database
Link and Scott (2012b)	This monograph presents a detailed description of the growth in employees from Phase II projects funded by the DOD, NIH, NASA, DOE, and NSF. These findings were later published in journal articles.	2005 Database
Link and Scott (2012c)	While firms generally hire additional employees to pursue Phase II projects, the authors quantify empirically that the number of employees retained after the completion of Phase II DOD, NIH, NASA, DOE, and NSF projects is small.	2005 Database
Link and Scott (2012d)	The authors explore quantitatively whether there is evidence that strategic commercial agreements allow foreign firms to exploit the technology developed though the SBIR programs at DOD, NIH, NASA, DOE, or NSF. The authors conclude that the evidence is minimal.	2005 Database
Audretsch, Leyden, and Link (2013)	The authors found that university involvement in DOE-funded Phase II projects is related to whether the researching firm engages in strategic business activities with other firms, thus taking advantage of knowledge spillovers.	2005 Database
Gicheva and Link (2013)	The authors conclude from an analysis of NIH Phase II projects that women-owned firms are at a disadvantage in attracting external development funding to bridge the gap from technology to innovation.	2005 Database

Author(s)	Key Findings	NRC Data
Gicheva and Link (2015)	Using Phase II data on projects funded by DOD, NIH, NASA, DOE, and NSF, the authors find that female-owned firms are disadvantaged in their access to private investment to assist in the development of their Phase II technology.	2005 Database
Link (2015)	Using Phase II data from projects funded by DOD, NIH, NASA, DOE, and NSF, the author examines covariates associated with universities as a research partner in the development of a Phase II technology.	2005 Database
Link and Wright (2015)	Based on information on Phase II projects funded by DOD, NIH, NASA, DOE, and NSF, the probability of the research project failing is reduced if the researching firm has prior research experience with the R&D projects similar in scope to the funded technology. The probability of failure also decreases the greater the project's R&D budget.	2005 Database
Audretsch, Kuratko, and Link (2016)	The authors conclude from an analysis of DOD, NIH, NASA, DOE, and NSF Phase II projects that academic-based human capital is positively correlated with dynamic behavior (i.e., successful innovative behavior), whereas as prior business experience is not.	2005 Database
Gicheva and Link (2016)	Using Phase II data from projects funded by the DOD, NIH, NASA, DOE, and NSF, the authors find that nascent firms are more likely to fail in their SBIR-supported R&D endeavors, but nascent firms that do not fail have a higher probability of commercializing their developed technology compared to more established firms.	2005 Database
Andersen, Bray, and Link (2017)	The authors conclude from an analysis of NIH Phase II projects that firms that have a founder with a business background are less likely to have their funded projects fail.	2005 Database
Scott, Scott, and Link (2017)	The authors' analysis of DOD-funded Phase II projects suggests that the probability of small firms obtaining outside financing to support their R&D projects is greater given more complex commercial opportunities, defined as a greater number of different potential applications for a project's anticipated results, for their innovations.	Survey data collected as part of NRC's first-round assessment
Audretsch and Link (2018)	The authors' analysis of innovation capital from DOD, NIH, NASA, DOE, and NSF Phase II projects concludes that the innovation capital within the firm is correlated with the commercial success of the firms' Phase II technology.	2005 Database

Author(s)	Key Findings	NRC Data
Hayter and Link (2018)	The authors present information that more than 50% of firms' Phase II projects funded by DOD, NIH, NASA, DOE, and NSF publish aspects of their research. The authors' interpretation of the literature is that entrepreneurial firms use competitive publication logic to invalidate or preempt intellectual property protection, and firms use accretive publication logic to attract financial resources, enhance the reputation of in-house scientists, and obtain new technologies critical to the development of new products.	2005 Database
Link and Scott (2018a)	The authors test and confirm the Schumpeterian hypothesis about the effect of firm size on R&D output for a sample of R&D projects for R&D intensive firms that are small using Phase II data for projects funded by DOE.	2005 Database
Link and Scott (2018b)	The authors conclude from an analysis of NIH-funded Phase II projects that, relative to a counterfactual control group, NIH can be characterized as supporting, on average, the development of high commercialization risk technologies.	2014 Database
Audretsch and Link (2019b)	Using Phase II project data from the DOD, NIH, NASA, DOE, and NSF, the authors conclude that firms with more technical experience and sector experience have realized higher growth rates from their SBIR-funded research.	2005 Database
Audretsch, Link, and Van Hasselt (2019)	Using information on scientific publication resulting from Phase II projects funded by the DOD, NIH, NASA, DOE, and NSF, the authors conclude that technical human capital knowledge from the university spills over to the firm's project and begets (i.e., brings about) additional knowledge in the form of scientific papers submitted for publication.	2005 Database
Link and Morrison (2019)	The authors conclude from an analysis of Phase II projects funded by DOD, NIH, NASA, DOE, and NSF that the probability of commercialization of the new technology does not vary by either the minority status or gender status of the firm.	2005 Database
Link and Van Hasselt (2020)	The authors use data on Phase II projects funded by DOD, NIH, NASA, DOE, and NSF to investigate the R&D-to-patenting relationship. The conclusion from their study is that women-owned firms are associated with fewer patent applications than men-owned firms. Minority-owned firms are associated with more patent applications than non-minority-owned firms, but the estimate of the difference is not statistically significant.	2005 Database

Author(s)	Key Findings	NRC Data
Bednar, Gicheva, and Link (2021)	The authors conclude from an analysis of Phase II projects funded by NIH that the synergetic effect of female-owned firms that have a female PI leading their research project means these firms are more successful in commercializing their technology.	2005 Database and 2014 Database
Cunningham and Link (2021a)	The authors rely on Phase II data from DOD, NIH, NASA, DOE, and NSF to explore the behavior of firms that do not commercialize their technology. The authors find that firms that do not commercialize their newly developed SBIR-funded technology have a greater probability of selling their technology to another firm.	2005 Database
Link (2021)	DOD-, NIH-, NASA-, DOE-, and NSF-funded Phase II projects were studied to determine the extent to which new technology information was transferred from the researching firms through publications. Relevant covariates are the amount of the funded research project, university involvement in the project, the firms' history of SBIR funding, and the academic background of the firms' founders.	2005 Database
Link, Van Hasselt, and Vismara (2021)	The authors conclude from their analysis of Phase II project funded by DOD, NIH, NASA, and DOE that the probability of an actual or planned Initial Public Offering (IPO) is higher for firms that hold patents, compared to firms without patents.	2011 Database and 2014 Database

Note: A number of the studies using Phase II data from the 2005 Database have been compiled and are published in Link (2013).

SUMMARY AND WHAT FOLLOWS

There is a rich literature in economics and in the policy disciplines about the SBIR program. In this chapter we have provided context, through a summary of the extant literature, for the empirical relationships discussed in the following chapters. Also, the datasets used throughout the remaining chapters are described.

In Chapter 5, we describe the demographics of the firms that responded to the various rounds of surveys. Specifically, we characterize the Phase II awards in terms of the gender and race of the owners of the firms awarded and conducting the research. The data underlying the descriptive analysis in Chapter 5 will be used in later chapters.

NOTES

1. As background, a history of these agencies is at: DOE: https://www.energy.gov/management/office-management/operational-management/history/brief-history-department-energy, NIH: https://history.nih.gov/exhibits/history/index.html, DOD: http://libguides.lib.msu.edu/c.php?g=96617&p=626695, NASA: http://time.com/3964417/nasa-history-1958/ and https://history.nasa.gov/factsheet.htm, and NSF: https://www.nsf.gov/about/history/history-publications.jsp. Websites accessed May 9, 2022.
2. We do not focus on the Small Business Technology Transfer (STTR) program in this book. The Small Business Research and Development Enhancement Act of 1992 (Public Law 1025–64) amended the Act of 1982. Title II of this Act is cited as the Small Business Technology Transfer Act of 1992, and it established the STTR Pilot program. According to the Small Business Administration, the purpose of the STTR program is to "facilitate the transfer of technology developed by a research institution through the entrepreneurship of a small business concern." See Link (2023). See also https://www.sbir.gov/tutorials/program-basics/tutorial-1. Accessed May 9, 2022.
3. Care should be exercised when interpreting the survey responses based on how the adjective *random* is used by the National Academies to describe each database. A number of filters were imposed on the population of Phase II projects including the availability of known addresses of potential respondents. Thus, any generalizations from the descriptive statistics in the following chapters to all Phase II projects should be made with caution.
4. See https://www.sbir.gov/sbirsearch/award/all. Accessed February 16, 2022.
5. Not all of the Phase II project questions were answered by every firm.
6. Statistically ($p = .08$), a 10 percent increase in the number of employees in the firm when the Phase II proposal was submitted is associated with an 11 percent increase in the amount of the Phase II award.

5. Demographics of SBIR awardees

DEMOGRAPHICS OF AWARDEES

As discussed in Chapter 2, and as illustrated in Table 2.1, the SBIR program has a long-legislated history of encouraging women and minorities to participate in the innovation process. The purpose statement in the enabling legislation—the 1982 Act—was [emphasis added]:

> ... to foster and encourage participation by ... *minority* and *disadvantaged persons* in technological innovation.

A similar statement was in the 1992 reauthorization of the program:

> ... to improve the Federal Government's dissemination of information concerning the small business innovation research program, particularly with regard to program participation by *women-owned* small business concerns and by *socially and economically disadvantaged* small business concerns.

And the current statement of the SBIR program's goals include:

> ... [to] foster and encourage participation in innovation and entrepreneurship by *women* and *socially or economically disadvantaged* persons.

In this chapter, we document across the five agencies the percentages of Phase II awards to both women-owned firms and to minority-owned firms.[1] And we illustrate the trend in these percentages over the year 1992 through 2007, 2009, or 2010, depending on the end date of each agency's second-round study. These illustrated trends should be interpreted in terms of the following context. Specifically, the enabling legislation and its amendments do not define the phrases *to foster*, *to encourage*, or *to improve* from the above quoted enabling legislations' statements. If one interprets these phrases to mean that the percentage of Phase II awards to women-owned firms or to minority-owned firms is expected to increase over time, then the illustrations below might have probative value when interpreting the extent that the SBIR program is meeting its purposes or goals. Alternatively, one could interpret the above legislated statements to refer to Phase I awards because receipt of a Phase I award is a necessary, but not sufficient, condition for receipt of a Phase II award. However, an equally valid interpretation could be that the program itself, meaning a program manager, is expected to provide additional information and proposal support to women-owned and minority-owned firms so that these firms are more competitive for a research award. Thus, the remainder of this chapter should be viewed as a descriptive snapshot of Phase II award recipients rather than a policy assessment.

A TREND ANALYSIS OF AWARDS BY DEMOGRAPHICS

A description of the trends of the gender and minority status of firms' owners across agencies is presented in this section. Figures 5.1 through 5.5 illustrate the percentages of Phase II awards to women-owned and to minority-owned firms, by award year.[2]

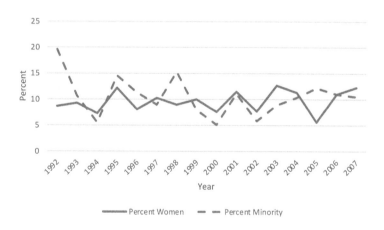

Figure 5.1   *DOD Phase II awards to women-owned and minority-owned firms, by award year*

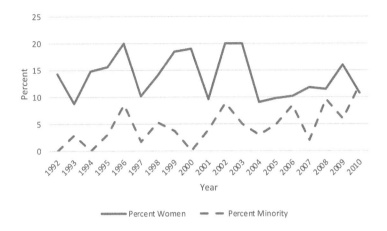

Figure 5.2   *NIH Phase II awards to women-owned and minority-owned firms, by award year*

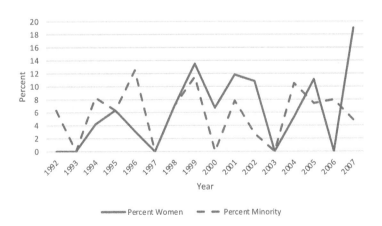

Figure 5.3   *NASA Phase II awards to women-owned and minority-owned firms, by award year*

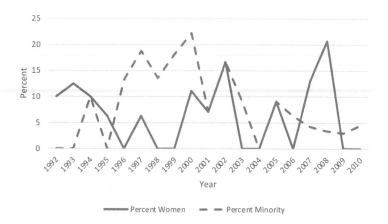

*Figure 5.4   DOE Phase II awards to women-owned and
                 minority-owned firms, by award year*

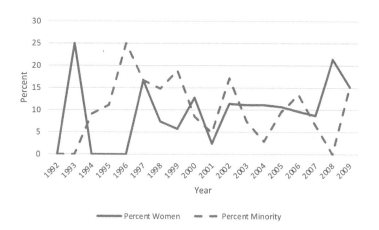

*Figure 5.5   NSF Phase II awards to women-owned and
                 minority-owned firms, by award year*

Several descriptive generalizations follow from these figures:

- Among DOD-funded projects, the percentage of awards to women-owned firms and to minority-owned firms has generally remained constant and approximately equal between the two ownership groups over time.
- Among NIH-funded projects, the percentage of awards to women-owned firms has been greater than to minority-owned firms, although a crossover occurred in 2010.
- Among NASA-funded projects, the percentage of awards to women-owned firms and to minority-owned firms has oscillated not only in terms of which percentage was greater year after year, but also in terms of the actual percentages.
- Among DOE-funded projects, there appears visually to be a slight upward trend among women-owned firms that were funded and a slight downward trend in the percentage of minority-owned firms that were funded.
- Among NSF-funded firms, the percentage of awards to women-owned firms and to minority-owned firms has oscillated not only in terms of which percentage was greater year after year but also in terms of the actual percentages.

SUMMARY AND WHAT FOLLOWS

The goals of the SBIR program that we discussed in Chapter 2 emphasize the importance of the participation of women and socially or economically disadvantaged persons in the innovation process. In this chapter, we have characterized the participatory behavior of firms owned by women and firms owned by minorities over time.

In Chapter 6, we describe alternative measures of what we call *productive capital* within the context of a performance production function. We define each measure of productive capital in terms of available information in the databases. The metrics underlying the descriptive analysis in Chapter 6 will be used in later chapters.

NOTES

1. We use the term *minority* to refer to socially or economically disadvantaged persons.
2. The number of observations used in these figures is less than the total number of projects in the combined dataset (see Table 4.1) because some firms were not responsive to the survey question about ownership.

6. Productive capital

ELEMENTS OF PRODUCTIVE CAPITAL

While many credit Hicks (1932) with the concept of a production function, especially a production function that relates to technological change, writings about production and distribution have a rich history that traces to the late 1700s as discussed by Humphrey (1997). Still, we begin our discussion with the Hicksian production function which can be written as:

$$Q = f(K, L) \tag{6.1}$$

where $f(\cdot)$ represents the production frontier; Q represents output, or in our case a measure of performance; K represents the stock of capital; and L represents the stock of labor. Over time, however, scholars have expanded their thoughts about the measurement of K and L and thus have introduced new methods to address the heterogeneity of the stock of these productive inputs. For example, various vintages of capital exist within a firm or economic unit and each vintage of capital could have a different level of productivity. Also, various units of labor have both different endowments of education and experience, and thus each cohort of labor could have a different level of productivity.

As production function analysis progressed over time, in large part due to the development of new specifications for $f(\cdot)$ above,[1] and due to the availably of more precisely measured inputs, the technology-based literature and the knowledge production literature began to consider alternatively defined variables as arguments in a production function.[2] Discussed below are four such internal productive capital variables: technical capital (TC), human capital (HC), experiential capital (EC), and social capital (SC).

In general terms, following Griliches (1979), the term *knowledge production function* refers to a transformation of various innovative inputs into various innovation outputs. Functionally, let IO represent an innovative output, and let Kn represent an innovative input which might be termed *knowledge*. Then:[3]

$$IO = f(Kn) \tag{6.2}$$

Knowledge within a firm or within an economic unit can be embodied in any one or all of the above-mentioned elements of productive capital. Thus:

$$Kn = G(TC, HC, EC, SC) \tag{6.3}$$

Or, based on equation (6.2), it follows that:

$$IO = F(TC, HC, EC, SC) \tag{6.4}$$

We offer a conceptual definition of each productive capital element below, and we describe the measurement of each subject to available information in the NRC databases in the discussion that follows. And we emphasize that these productive capital elements are all internal elements to the firm and definitionally are all present in the funded firm prior to the receipt of a Phase II research award.

Technical Capital (TC)

Conceptually, technical capital refers to the award amount of R&D available to the firm to perform the funded Phase II project. As previously noted, the Phase II award amounts were not provided in the 2011 Database and the 2014 Database for privacy considerations. That said, in a number of the previous studies of the performance of Phase II projects (see Table 4.2), the Phase II award amount was not a statistically significant covariate.

Human Capital (HC)

Conceptually, human capital refers to the knowledge embodied within individuals within a firm. The databases provide information on the number of employees in the firm either at the time the Phase II proposal was submitted or at the time the Phase II award was made. The nature of the survey question depends on the database in hand (i.e., slightly different questions for different surveys). We assume that the number of employees is approximately the same at the date of the Phase II proposal as it would be at the date of the award (or vice versa).

Unfortunately, there is not uniform information to allow us to weight employees based on their education level or their administrative or sectoral experiences. Thus, HC is measured here as the number of employees in the firm prior to the beginning of the Phase II research.[4]

Experiential Capital (EC)

SBIR proposals are peer reviewed, and a proposal element considered by reviewers is the likelihood of technical success of the proposed Phase II project and the likelihood of commercial success of the developed technology. As noted in Chapter 1, a current SBIR program goal is to "increase private-sector commercialization of innovations" (where, as noted in Table 3.1, a technology that enters the market is generally referred to as an innovation).

The prior success of the proposing firm in receiving a Phase II award is arguably one indicator of the research experience or research reputation of the firm, and thus this metric may be a proxy for the experiential capital that is embodied in a firm. Thus, EC is measured as the number of previous Phase II awards the firm received that are related to the project/technology supported by the current Phase II award.

Social Capital (SC)

The sociological origin of the concept of social capital arguably traces to Hanifan (1916, p. 130) where it is defined as "being a part of a larger group." After nearly five decades, the concept of social networks gained prominence in the sociology literature. In terms of social capital relating to entrepreneurial behavior, and small firms that participate in the SBIR program are often referred to as entrepreneurial firms based on their nascency and size, Granovetter (1973, p. 1361) argued that the entrepreneur's social network can be characterized as a mix of strong and weak ties with others, the distinction between those ties depending on "the amount of time, the emotional intensity, the intimacy (mutual confiding), and the reciprocal services which characterize the tie."[5]

Using the network relationship that characterizes social capital, SC is measured dichotomously in terms of whether (= 1) or not (= 0) the firm received from private-sector entities (e.g., venture capitalist, angel investors, other private-sector firms) prior to its current Phase II award any

R&D funding related to the technology being researched in the current Phase II project.[6]

The above operational definitions of productive capital that are supported by data from the NRC databases are summarized in Table 6.1.

Table 6.1 Definitions of internal elements of productive capital supported by data from the NRC databases

Elements	Operational Definition
Human Capital (HC)	HC is measured as the *number of employees* in the firm prior to the beginning of the Phase II research.
Experiential Capital (EC)	EC is measured as the *number of previous Phase II awards* the firm received that are related to the project/technology supported by the current Phase II award.
Social Capital (SC)	SC is measured dichotomously in terms of whether (= 1) or not (= 0) the firm received *additional R&D funding* from private-sector entities (e.g., venture capitalist, angel investors, other private-sector firms) prior to its Phase II award related to the technology being researched in the Phase II project.

QUANTIFYING ELEMENTS OF PRODUCTIVE CAPITAL

Table 6.2 shows the mean values of the three quantifiable productive capital elements, by agency. For the construction of these tables, if a firm was missing values for any of the capital elements or for gender and minority status, the Phase II project was removed from the sample.[7] Thus, the number of observations (n) for each element of productive capital are the same, by agency.[8]

Table 6.2   *Descriptive statistics on internal elements of productive capital, by funding agency*

Funding Agency/Productive Capital Element	Mean	Standard deviation	Minimum	Maximum
DOD (n = 1,463)				
Number of employees (HC)	36.35	61.76	1	490
Number of previous awards (EC)	1.17	1.73	0	28
Additional funding (SC)	0.15	0.36	0	1
NIH (n = 820)				
Number of employees (HC)	21.99	46.81	1	422
Number of previous awards (EC)	1.23	2.63	0	28
Additional funding (SC)	0.12	0.32	0	1
NASA (n = 306)				
Number of employees (HC)	39.35	63.61	1	375
Number of previous awards (EC)	1.17	1.67	0	12
Additional funding (SC)	0.11	0.32	0	1
DOE (n = 315)				
Number of employees (HC)	34.30	54.08	1	450
Number of previous awards (EC)	1.07	1.73	0	15
Additional funding (SC)	0.08	0.27	0	1
NSF (n = 316)				
Number of employees (HC)	18.84	32.21	1	200
Number of previous awards (EC)	1.26	2.57	0	40
Additional funding (SC)	0.28	0.45	0	1

Tables 6.3 through 6.7 present a correlation matrix of the productive capital elements as well as the binary variables for women-owned and minority-owned firms.

Table 6.3   *Correlations among DOD-funded firms' internal elements of productive capital, n = 1,463*

	Award year	Women-owned	Minority-owned	Number of employees (HC)	Number of previous awards (EC)	Additional funding (SC)
Award year	1					
Women-owned	0.025	1				
Minority-owned	−0.005	−0.018	1			
Number of employees (HC)	−0.011	−0.117***	−0.085***	1		
Number of previous awards (EC)	0.152***	−0.069***	−0.042*	−0.008	1	
Additional funding (SC)	0.042	0.009	−0.026	−0.082***	0.104***	1

Note: *** significant at the .01-level, * significant at the .10-level.

Table 6.4   *Correlations among NIH-funded firms' internal elements of productive capital, n = 820*

	Award year	Women-owned	Minority-owned	Number of employees (HC)	Number of previous awards (EC)	Additional funding (SC)
Award year	1					
Women-owned	−0.027	1				
Minority-owned	0.103***	0.046	1			
Number of employees (HC)	−0.040	−0.027	0.047	1		
Number of previous awards (EC)	0.006	0.186***	−0.008	0.087***	1	
Additional funding (SC)	−0.203***	−0.070**	−0.054	−0.030	−0.073**	1

Note: *** significant at the .01-level, ** significant at the .05-level.

Table 6.5   Correlations among NASA-funded firms' internal elements of productive capital, n = 306

	Award year	Women-owned	Minority-owned	Number of employees (HC)	Number of previous awards (EC)	Additional funding (SC)
Award year	1					
Women-owned	0.098*	1				
Minority-owned	−0.082	−0.081	1			
Number of employees (HC)	−0.055	−0.080	−0.135**	1		
Number of previous awards (EC)	0.159***	0.019	0.061	−0.042	1	
Additional funding (SC)	0.111**	0.075	−0.053	−0.052	0.192***	1

Note: *** significant at the .01-level, ** significant at the .05-level, * significant at the .10-level.

Table 6.6   Correlations among DOE-funded firms' internal elements of productive capital, n = 315

	Award year	Women-owned	Minority-owned	Number of employees (HC)	Number of previous awards (EC)	Additional funding (SC)
Award year	1					
Women-owned	−0.001	1				
Minority-owned	−0.071	0.030	1			
Number of employees (HC)	0.077	−0.082	−0.007	1		
Number of previous awards (EC)	0.033	−0.083	−0.005	−0.014	1	
Additional funding (SC)	−0.208***	0.034	−0.045	−0.053	0.043	1

Note: *** significant at the .01-level.

Table 6.7   *Correlations among NSF-funded firms' internal elements of productive capital, n = 316*

	Award year	Women-owned	Minority-owned	Number of employees (HC)	Number of previous awards (EC)	Additional funding (SC)
Award year	1					
Women-owned	0.097*	1				
Minority-owned	0.007	0.123**	1			
Number of employees (HC)	−0.089	−0.068	−0.002	1		
Number of previous awards (EC)	0.145***	−0.036	−0.052	0.084	1	
Additional funding (SC)	0.033	−0.049	−0.026	0.065	0.112**	1

Note: *** significant at the .01-level, ** significant at the .05-level, * significant at the .10-level.

We offer several conclusions across agencies from the descriptive information in these matrices. These are as follows:

- There is no consistent trend in ownership over the years across the agencies. We do observe that the number of minority-owned NIH-funded firms has increased over the award years (correlation: 0.103), and the number of women-owned NASA-funded firms and NSF-funded firms has increased over the award years, based on positive and statistically significant correlation coefficients (0.098 and 0.097 for NASA and NSF, respectively).[9]
- Across agencies, there is no time trend in the human capital measure of productive capital. The correlation coefficients between award years and human capital are all not statistically significant.
- Measures of experiential capital have increased over time among DOD-, NASA-, and NSF-funded firms (correlations: 0.152, 0.159, and 0.145, respectively). One might speculate that this finding implies a legacy pattern in the receipt of Phase II awards.
- Statistically significant correlations between the social capital measure (i.e., the presence of external funding for the funded Phase II research) and award year were found for firms funded by NIH, NASA, and DOE. The signs of the correlations, however, varied.

One might speculate that social capital endowments among firms is an agency effect rather than a predictable pattern of funding over time.

- Across agencies, human capital endowments are generally smaller among women-owned and minority-owned firms based on a negative correlation coefficient. The correlations among these variables are insignificant, except in the case of DOD and NASA for minority-owned firms.
- There is no consistent relationship across agencies between experiential capital or social capital endowments and women or minority ownership status of funded firms. With the exception of DOD and NIH, the relevant correlation coefficients are not statistically significant. For DOD and NIH, some of the correlations are statistically significant but they vary in algebraic sign.
- Except for NIH- and DOE-funded firms, the correlation between social capital and experiential capital is positive and statistically significant. One possible explanation for this is that prior success in obtaining SBIR funding might help a firm in securing external funding for their current Phase II research projects.

SUMMARY AND WHAT FOLLOWS

Three elements of productive capital within SBIR-funded firms were defined in this chapter, and the inter-relationship of these elements was shown through correlation analysis. The elements are human capital, experiential capital, and social capital.

In Chapter 7, descriptive information is presented about the failure of funded Phase II projects. Failure is defined in terms of a Phase II project being discontinued before completion, and reported reasons for being discontinued are presented. Also, project failure is shown to be related to the presence of productive capital. The analysis that follows Chapter 7 is based on Phase II projects that did not fail, and the productive capital elements will be revisited.

NOTES

1.   In the context of technological change, see the discussion of production functions in Link and Siegel (2003).
2.   Versions of equation (6.1) have been used to quantify residually measured technological change (Solow, 1957) and the returns to private and public investments in R&D (e.g., Terleckyj, 1974; Link and Scott, 2020, 2021).

3. This functional form is discussed in Link and Van Hasselt (2020).
4. To the extent to which a firm used a portion of its Phase II award to hire additional employees, HC underestimates the amount of human capital in the firm during the Phase II research.
5. Leyden and Link (2015a, 2015b) have developed this concept of strong and weak ties within an entrepreneur's network into their theory of the entrepreneurial process.
6. In Chapter 10, we use this measure of social capital to define the presence of financial stakeholders in a Phase II research project.
7. The number of observations in Table 6.2, by agency, is less than the number of observations for gender and minority status of firm owners in Figures 5.1 through 5.5.
8. We believe that having a balanced panel is important for an interpretative comparison of the correlation coefficients below.
9. We are using here the number of firms that are minority-owned and woman-owned, whereas in Chapter 5 we looked at the trend in the percentage of Phase II awards to minority-owned and to woman-owned firms.

7. When research fails

A MEASURE OF FAILED RESEARCH

Phase I research awards are intended to establish the technical merit, feasibility, and commercial potential of the proposed research, and to determine the quality of performance of the small business awardee's organization prior to providing further federal support in Phase II. As so described, Phase I research might be thought of as a demonstration of proof of concept.

Not all firms that complete their Phase I research project apply for a Phase II award. Those firms that do receive a Phase II award have essentially passed a rigorous peer review process or, stated differently, have cleared a research hurdle. Nevertheless, some Phase II projects are discontinued before the proposed research and/or the commercialization effort is completed. Receiving a Phase II research award is therefore not a sufficient condition for ensuring that the research project will be successful.

The survey instruments that supported the databases discussed in Chapter 4 contain a question about the current status of the project funded by the SBIR award on which the survey is focused. The response statement (*Yes* or *No*) that forms the definition of project failure in this chapter is common across all databases:

> Efforts at this company have been discontinued. No sales or additional funding resulted from this project.

Our interpretation of a *Yes* response is that the Phase II research failed to meet its research goals or failed to commercialize the resulting technology. Table 7.1 shows the percentages of Phase II projects that failed by this definition, by funding agency.

Table 7.1   *Percent of Phase II projects that failed, by funding agency*

Funding agency	Number of projects	Percent of failed projects
DOD	1,463	22.3%
NIH	820	19.4%
NASA	306	23.2%
DOE	315	21.0%
NSF	316	16.1%

Also, Figures 7.1 through 7.5 show the percentage of Phase II projects that failed, by year and by funding agency. An inspection of the trend lines imposed on these figures suggests that project failure has generally been declining over time, or at least over recent periods of time.

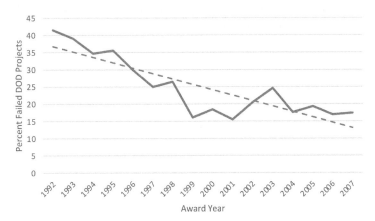

Note: The dashed line is a linear trend line.

Figure 7.1   *Percent of Phase II DOD-funded projects that failed, by award year, n = 1,463*

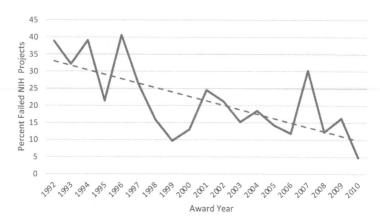

Note: The dashed line is a linear trend line.

Figure 7.2   *Percent of Phase II NIH-funded projects that failed, by award year, n = 820*

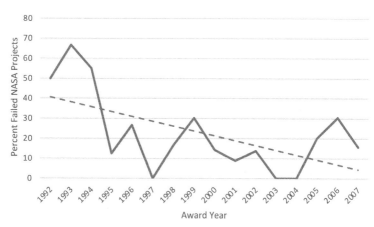

Note: The dashed line is a linear trend line.

Figure 7.3   *Percent of Phase II NASA-funded projects that failed, by award year, n = 306*

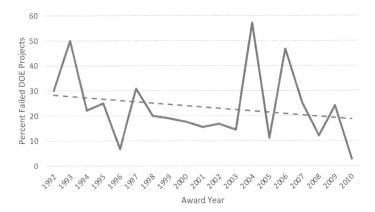

Note: The dashed line is a linear trend line.

Figure 7.4   *Percent of Phase II DOE-funded projects that failed, by award year, n = 315*

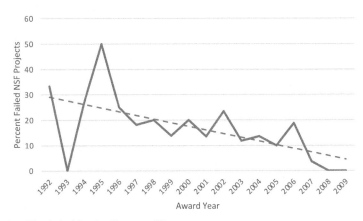

Note: The dashed line is a linear trend line.

Figure 7.5   *Percent of Phase II NSF-funded projects that failed, by award year, n = 316*

There are few academic studies about project failure. The empirical side of that research is even more limited and characterized predominantly by the studies of Link and Wright (2015), Andersen et al. (2017), and Link et al. (2022). As a starting point to summarize this body of literature, we point out the following opinion statement, which might temper one's interpretation of the findings we present below: "The concept of project failure is nebulous" (Pinto and Mantel, 1990, p. 269).

Shepherd and Wiklund (2006) ably reviewed the so-called project failure literature, although their arguments are not specific to research projects and certainly not specific to SBIR projects. Their review, however, represents an important starting point to motivate the analysis presented in this chapter. These authors offer three reasons for project failure.

- First, projects may fail because of insufficient human capital. This argument for failure traces to Becker (1975); Lazear (2005) has formalized the relationship between human capital and successful entrepreneurial behavior.
- Second, projects fail because of a liability of newness: "The liability of newness [i.e., lack of experience] relates to the actions and learning that the management team and employees must undergo to overcome the major challenges of adaptation to the internal and external environments of new organizations" (Shepherd and Wiklund, 2006, p. 5).[1]
- Third, projects may fail because of overconfidence. Overconfidence results from hubris, and hubris leads to a tendency to deprive projects with needed resources and resourcefulness.[2] The lack of resources will then increase the likelihood of failure.

Our interpretation of the above reasons for project failure is that probability of failure decreases (increases) as the human capital, the experience base of either the project manager or firm, or the social capital of the firm increases (decreases), other factors remaining constant. Therefore, we anticipate that the elements of productive capital discussed in the previous chapter—human capital, experiential capital, and social capital—are likely to be negatively correlated with the likelihood of an SBIR Phase II project failing.

EMPIRICAL FINDINGS

As noted in Table 7.1, the fraction of projects that fail ranges across agencies from 16.1 percent to 23.2 percent.

There are several patterns among the correlations of productive capital measures and project failure that are noticeable in Tables 7.2 through 7.6. First, the correlation between the award year and project failure is negative in all funding agency samples of projects, and it is significant for all agencies with the exception of the DOE. The negative relationship is consistent with the patterns shown in Figures 7.1 through 7.5. An inspection of Figure 7.4 shows the irregular pattern of the fraction of failed projects across years, sporadic nature of the average probability of project failure across years, so the insignificance of the correlation among the DOE sample of projects might not be surprising.

Second, the correlation between project failure and women or minority ownership status of firms is not significant across any of the funding agency samples of projects. Gender and minority status of the owners of wThird, as suggested through our interpretation of the literature above,

Table 7.2   *Correlations among DOD-funded firms' internal elements of productive capital and project failure, n = 1,463*

	Award year	Women-owned	Minority-owned	Number of employees (HC)	Number of previous awards (EC)	Additional funding (SC)	Project failure
Award Year	1						
Women-owned	0.025	1					
Minority-owned	−0.005	−0.018	1				
Number of employees (HC)	−0.011	−0.117***	−0.085***	1			
Number of previous awards (EC)	0.152***	−0.069***	−0.042*	−0.008	1		
Additional funding (SC)	0.042	0.009	−0.026	−0.082***	0.104***	1	
Project failure	−0.147***	−0.040	0.025	0.099***	−0.157***	−0.074***	1

Note: *** significant at the .01-level, * significant at the .10-level.

Table 7.3 *Correlations among NIH-funded firms' internal elements of productive capital and project failure, n = 820*

	Award year	Women-owned	Minority-owned	Number of employees (HC)	Number of previous awards (EC)	Additional funding (SC)	Project failure
Award Year	1						
Women-owned	−0.027	1					
Minority-owned	0.103***	0.046	1				
Number of employees (HC)	−0.040	−0.027	0.047	1			
Number of previous awards (EC)	0.006	0.186***	−0.008	0.087***	1		
Additional funding (SC)	−0.203***	−0.070**	−0.054	−0.030	−0.073**	1	
Project failure	−0.144***	0.028	−0.021	0.077**	−0.101***	0.002	1

Note: *** significant at the .01-level, ** significant at the .05-level.

Table 7.4 *Correlations among NASA-funded firms' internal elements of productive capital and project failure, n = 306*

	Award year	Women-owned	Minority-owned	Number of employees (HC)	Number of previous awards (EC)	Additional funding (SC)	Project failure
Award Year	1						
Women-owned	0.098*	1					
Minority-owned	−0.082	−0.081	1				
Number of employees (HC)	−0.055	−0.080	−0.135**	1			
Number of previous awards (EC)	0.159***	0.019	0.061	−0.042	1		
Additional funding (SC)	0.111**	0.075	−0.053	−0.052	0.192***	1	
Project failure	−0.204***	−0/029	−0.020	0.022	−0.097*	−0.052	1

Note: *** significant at the .01-level, ** significant at the .05-level, * significant at the .10-level.

Table 7.5 *Correlations among DOE-funded firms' internal elements of productive capital and project failure, n = 315*

	Award year	Women-owned	Minority-owned	Number of employees (HC)	Number of previous awards (EC)	Additional funding (SC)	Project failure
Award Year	1						
Women-owned	−0.001	1					
Minority-owned	−0.071	0.030	1				
Number of employees (HC)	0.077	−0.082	−0.007	1			
Number of previous awards (EC)	0.033	−0.083	−0.005	−0.014	1		
Additional funding (SC)	−0.208***	0.034	−0.045	−0.053	0.043	1	
Project failure	−0.075	−0.019	−0.013	0.175***	−0.198***	−0.036	1

Note: *** significant at the .01-level,

Table 7.6 *Correlations among NSF-funded firms' internal elements of productive capital and project failure, n = 316*

	Award year	Women-owned	Minority-owned	Number of employees (HC)	Number of previous awards (EC)	Additional funding (SC)	Project failure
Award Year	1						
Women-owned	0.097*	1					
Minority-owned	0.007	0.123**	1				
Number of employees (HC)	−0.089	−0.068	−0.002	1			
Number of previous awards (EC)	0.145***	−0.036	−0.052	0.084	1		
Additional funding (SC)	0.033	−0.049	−0.026	0.065	0.112**	1	
Project failure	−0.143***	−0.080	0.001	0.148***	−0.007	−0.116**	1

Note: *** significant at the .01-level, ** significant at the .05-level, * significant at the .10-level.

firms endowed with greater human capital should be less likely to fail, so the correlation coefficients between these two variables should be negative. However, perhaps due to the lack of precision of human capital as measured by the funded firm's number of employees, the human capital variable is *positively*, not negatively, related to project failure. Among the project samples for DOD, NIH, DOE, and NSF, the correlations between human capital and project failure are also significant. A potential explanation of this finding might be that larger-sized firms with greater endowments of human capital are more likely to be overconfident in their research abilities and thus are less precise in their technology development and related market analysis.

Fourth, in concert with our interpretation of the literature, experiential capital is negatively correlated with project failure in all funding agency samples of projects, and the correlation is significant for all funding agencies except NSF.

Fifth, social capital is negatively correlated with project failure in all agency samples of projects, with the exception of the NIH sample. The correlation is negative and significant only in the DOD sample and the NSF sample. We revisit the role of social capital in Chapter 10. There we offer an interpretation of social capital as a measure of the presence of stakeholders in the success of the Phase II project, and we find that stakeholders and their resources do matter. When stakeholders have been involved with firms in their technology development of the currently funded SBIR Phase II project, the firms are more likely to be successful as measured in terms of their commercialization of their Phase II developed technology.

To summarize, the correlation analysis in the tables above is consistent in there being a decrease in project failure over time. The gender or minority status of the firm's owner does not appear to be related to project failure. Also, larger firms—as measured by the number of employees, with larger firms having a greater endowment of human capital—have a greater number of failed Phase II research projects. Firms with greater endowments of experiential capital have fewer failed Phase II projects. And finally, firms that have social capital—as measured by having had previous investors in the currently funded Phase II technology—have lower failure rates.

WHY PHASE II PROJECTS FAIL

As noted above, the survey question that we used to define project failure is as follows:

> Efforts at this company have been discontinued. No sales or additional funding resulted from this project.

In addition to this survey question, respondents were asked about the *primary* reason for discontinuing the Phase II project. As has previously been the case, the choices given to respondents varied across agency survey instruments, but there were seven responses that were common to all surveys and thus to all databases. These response categories for the primary reason for discontinuing the Phase II project are:

- Technical failure or difficulties
- Market demand too small
- Level of technical risk too high
- Company shifted priorities
- Principal investigator left
- Licensed (technology) to another company
- Inadequate sales capabilities.

Table 7.7 shows the percentage of responses to each of the primary reasons listed above by funding agency. The reported reasons for a Phase II project being discontinued vary across funding agencies. For all agencies, the top two primary reasons for failure are market demand being too small and technical failure or difficulties. Except for NIH, the top primary reason was that market demand was too small.

Table 7.7　　　　*Percent of Phase II projects that failed, by primary reason for being discontinued and by funding agency*

Funding agency (n)	Technical failure or difficulties (%)	Market demand too small (%)	Level of technical risk too high (%)	Company shifted priorities (%)	Principal investigator left (%)	Licensed (technology) to another company (%)	Inadequate sales capabilities (%)
DOD (n = 1,463)	3.1	6.6	1.0	1.6	0.4	0.8	1.23
NIH (n = 820)	4.9	4.8	1.2	5.6	0.7	1.7	1.5
NASA (n = 306)	3.9	8.5	1.3	1.6	1.3	0.7	1.6
DOE (n = 315)	7.9	8.9	2.2	4.4	1.0	1.3	1.6
NSF (n = 316)	9.8	13.0	4.4	8.9	3.8	5.7	3.5

Not only do the reasons for a Phase II project being discontinued vary across agencies, but also, they vary over time for each agency. Figures 7.6 through 7.10 show the percentage of responses for three of the primary reasons listed above, by funding agency and by fiscal year. The primary reasons shown in these figures are technical failure or difficulties, market demand too small, and level of technical risk too high. Caution should be exercised in generalizing from the NASA, DOE, and NSF samples because of a small number of projects in each award year for which a primary reason was reported by the survey respondent. That said, one generalization that can be suggested is that the proportions of Phase II projects being discontinued for these three primary reasons have generally, although not always smoothly, decreased over time.

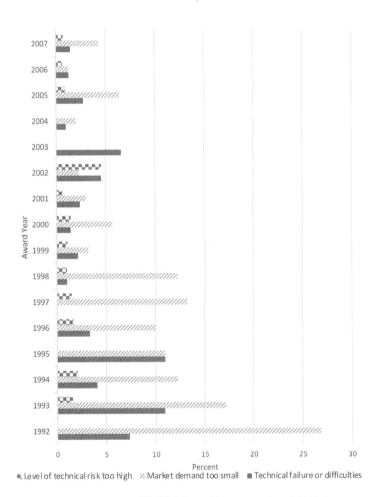

*Figure 7.6   Percent of DOD Phase II projects that failed, by
primary reason for being discontinued and by award
year, n = 1,463*

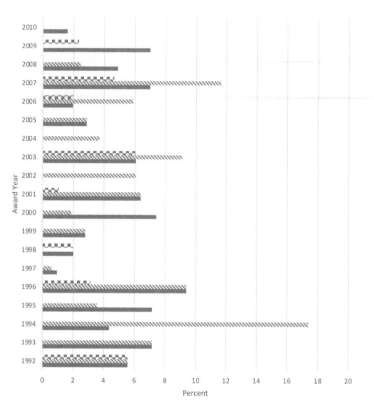

Figure 7.7   *Percent of NIH Phase II projects that failed, by primary reason for being discontinued and by award year, n = 820*

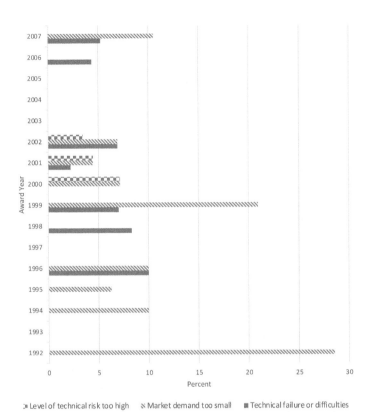

Figure 7.8   *Percent of NASA Phase II projects that failed, by primary reason for being discontinued and by award year, n = 306*

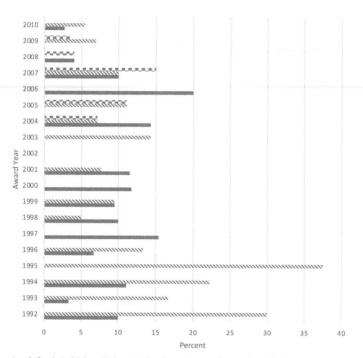

Figure 7.9   *Percent of DOE Phase II projects that failed, by primary reason for being discontinued and by award year, n = 315*

SUMMARY AND WHAT FOLLOWS

The data show that some Phase II projects have failed. In this chapter, we showed that the proportion of Phase II projects that fail has been decreasing over time. The likelihood of project failure is inversely related to the presence of experiential and social capital within the firm. Interestingly, larger endowments of human capital, as measured by the number of employees at the time of the Phase II award, are associated with an increased risk of project failure.

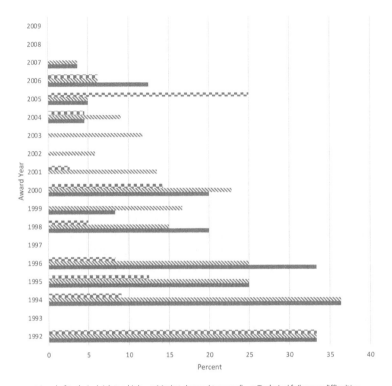

Figure 7.10   *Percent of NSF Phase II projects that failed, by primary reason for being discontinued and by award year, n = 316*

In Chapter 8, we evaluate to what extent SBIR-funded firms used external productive capital in their Phase II research projects. From the available information in the NRC databases, we quantify external productive capital through a binary indicator for the availability of university resources in the Phase II research project.

NOTES

1.   Nixon et al. (2012) embrace this concept as they argue that leadership is critical for project success, and newness can be synonymous with a lack of leadership.
2.   This argument follows from the Hayward et al. (2006) hubris theory of entrepreneurship.

8. University collaborators on SBIR projects

DEFINING UNIVERSITY COLLABORATION

The 2005 Database, the 2011 Database, and the 2014 Database contain response information to survey questions related to universities being collaborators on Phase II SBIR projects. The survey question about the involvement of university resources varied somewhat between the first round and second round of data collection. For example, the survey question in the first round of data collection was the same for all five agencies used in the construction of the 2005 Database. The *Yes/No* survey question was:

> In executing this award, was there any involvement by university faculty, graduate students, and/or university developed technologies?

The survey question immediately following this asked more specifically about the use of faculty, graduate students, and developed technologies. In the second round of data collection, the above broad question was not asked but only the question about the specific use of university resources. In what follows in this chapter, responses to the second-round survey question about university involvement were aggregated into a binary variable to mirror the *Yes/No* survey question from the first round of data collection.

We refer to the *involvement of university* resources of any type or in any form as an element of external productive capital. Table 8.1 shows the percentage of Phase II projects that involved a university or, more precisely, that involved university resources. Phase II projects funded by NIH and NSF relied on this form of external productive capital, which we refer to here as university capital (UC), to a greater extent than Phase II projects funded by the other three funding agencies.

Table 8.1   *Percent of Phase II projects that involved university resources, by funding agency*

Funding agency	Total number of projects	Projects with university involvement
DOD	1,463	30.5%
NIH	820	58.8%
NASA	306	31.7%
DOE	315	41.0%
NSF	316	57.3%

TRENDS IN UNIVERSITY COLLABORATION

Across funding agencies, university involvement has changed over time. Figures 8.1 through 8.5 show, for each agency, the percentage of Phase II projects that involved university resources by award year. For DOD (see Figure 8.1), aside from the dip in university involvement in 2007, the trend is generally positive. The same appears to be the case for NIH, although the estimated trend line (see the dashed line in Figure 8.2) is less steep than for DOD. Keeping in mind that the sample sizes decrease significantly among the NASA-, DOE-, and NSF-funded projects, there are notable periods of a positive trend that might be visually lost because of selected years (with a small number of sample projects) in which the percent of projects involving a university are zero. This is the case specifically among NSF-funded Phase II projects. If one were to ignore the award years 2008 and 2009, the trend among NSF-funded Phase II projects would also be positive.

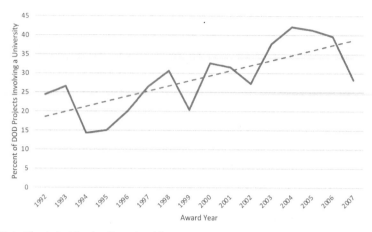

Note: The dashed line is a linear trend line.

Figure 8.1   *Percent of Phase II DOD-funded projects that involved*
                    university resources, by award year, n = 1,463

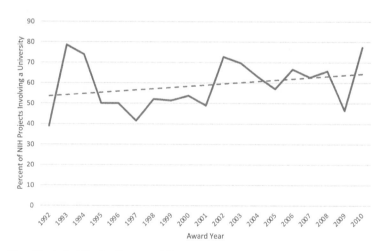

Note: The dashed line is a linear trend line.

Figure 8.2   *Percent of Phase II NIH-funded projects that involved*
                    university resources, by award year, n = 820

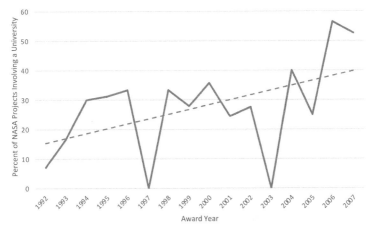

Note: The dashed line is a linear trend line.

Figure 8.3   *Percent of Phase II NASA-funded projects that involved university resources, by award year, n = 306*

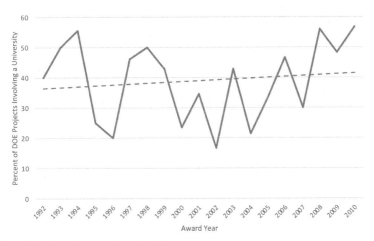

Note: The dashed line is a linear trend line.

Figure 8.4   *Percent of Phase II DOE-funded projects that involved university resources, by award year, n = 315*

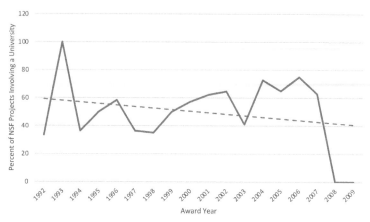

Note: The dashed line is a linear trend line.

Figure 8.5   Percent of Phase II NSF-funded projects that involved university resources, by award year, n = 316

The upward trend or periods of an upward trend are perhaps not surprising when one reflects on the findings from the 2008 report of the President's Council of Advisors on Science and Technology (PCAST), titled *University-Private Sector Research Partnerships in the Innovation Ecosystem.* Therein it is stated (PCAST, 2008, p. 19):

> Private sector engagement with researchers in academic … laboratories is increasingly vital to the health of U.S. R&D, and ultimately to the technology-based economy.

Among the justifications for this statement, PCAST appropriately noted that (2008, p. 19):

> This is because … industrial basic research laboratories have been reduced in both number and size and therefore industry has come to rely further on academic as well as government laboratories for basic research output … and the escalating pace of technology development calls for enhanced and novel technology transfer processes to capture these developments.

In other words (2008, p. 27):

> Universities continue to serve as a primary engine for discovery research that can lead to innovation ...

UNIVERSITY COLLABORATION AND PROJECT FAILURE

Our interpretation of the findings in the PCAST report is that universities can provide relevant resources to enhance or leverage the researching firm's own resources. Put somewhat differently, universities can provide needed research resources when they are unavailable within the researching firm itself. It follows from this interpretation that collaborations with universities might decrease the likelihood that the SBIR-funded Phase II project will fail.

Tables 8.2 through 8.6 show, by agency, the correlations between university involvement (0 or 1), the award year, and (the indicator for) project failure. There is a positive and significant correlation between university involvement and award year in all of the agencies. Also, there is a negative and significant correlation, as hypothesized, between university involvement and project failure among the DOD, DOE, and NSF projects, although the level of significant among the NSF projects is marginal ($p = .11$).[1]

Table 8.2   *Correlations among DOD's university productive capital and project failure, n = 1,463*

	Award year	Project failure	University involvement (UC)
Award year	1		
Project failure	−0.147***	1	
University involvement (UC)	0.124***	−0.055**	1

Note: *** significant at the .01-level, ** significant at the .05-level.

Table 8.3   *Correlations among NIH's university productive capital and project failure, n = 820*

	Award year	Project failure	University involvement (UC)
Award year	1		
Project failure	−0.144***	1	
University involvement (UC)	0.089***	0.011	1

Note: *** significant at the .01-level.

Table 8.4   *Correlations among NASA's university productive capital and project failure, n = 306*

	Award year	Project failure	University involvement (UC)
Award year	1		
Project failure	−0.204***	1	
University involvement (UC)	0.149***	0.091	1

Note: *** significant at the .01-level.

Table 8.5   *Correlations among DOE's university productive capital and project failure, n = 315*

	Award year	Project failure	University involvement (UC)
Award year	1		
Project failure	−0.075	1	
University involvement (UC)	0.101*	−0.096*	1

Note: * significant at the .10-level.

Table 8.6   *Correlations among NSF's university productive capital and project failure, n = 316*

	Award Year	Project failure	University involvement (UC)
Award year	1		
Project failure	−0.143***	1	
University involvement (UC)	0.139***	−0.091	1

Note: *** significant at the .01-level.

SUMMARY AND WHAT FOLLOWS

Some firms use university resources as complements to their internal productive capital. The data might be interpreted to show that when university resources are employed, the likelihood that the Phase II research will fail decreases. But, as noted, there is likely more than one reason that a university is invited to participate in a Phase II project, and that fact complicates the interpretation of our findings.

In Chapter 9, we investigate the relationship between knowledge production, as measured in terms of scientific publications and patents resulting from the Phase II projects, and the various elements of productive capital.

NOTE

1. It might also be the case that projects that involve university resources are also more complex (hence, the need for university involvement), which may ultimately make them more risky and sometimes more likely to fail.

9. Knowledge production functions

INTRODUCTION

In this chapter, we focus on two dimensions of the knowledge that is produced from the Phase II research process: patents that are related to the underlying technology that resulted from, or is resulting from, the Phase II research, and related scientific publications. Herein, we investigate the relationship between the elements of productive capital defined in Chapter 6 and Chapter 8 and these two knowledge outputs.[1] Patents received and scientific publications are the second and third categories (the first being policy transfers) of social benefits associated with the SBIR program that are described in this book.[2]

Our starting point is a historical trace of ideas about the sources of knowledge. It follows from this historical perspective that there is a tractable, as well as conceptual, relationship between one's experiences—the sources of knowledge that one relies on—and one's entrepreneurial actions. Perhaps, then, there is a relationship between a firm's experiences and its knowledge outputs.

John Locke was born in 1632. Educated in medicine at the University of Oxford, Locke soon transcended his formal training to become one of the most influential philosophers of his time, earning, posthumously, the titles of Founder of British Empiricism and Father of Classical Liberalism.[3] His most famous treatise is *An Essay Concerning Human Understanding*, first published in 1689 (dated 1690). Locke emphasized therein that all ideas emanate from sensation or reflection (Locke, 1996, p. 33; original emphasis throughout):

> Every man being conscious to himself, that he thinks, and that which his mind is employed about whilst thinking being the *ideas*, that are there, 'tis past doubt, that men have in their minds several *ideas … All ideas come from sensation or reflection.*

Regarding sensation, Locke emphasized one's perception of things (Locke, 1996, pp. 33–4):

> [O]ur senses, conversant about particular sensible objects, do *convey into the mind* several distinct *perceptions* of things, according to those various ways, wherein those objects do affect them: and thus we come by those *ideas* we have of *yellow, white, heat, cold, soft, hard, bitter, sweet*, and all those which we call sensible qualities, which when I say the senses convey into the mind, I mean, they from external objects convey into the mind what produces there those *perceptions*. This great source, of most of the *ideas* we have, depending wholly upon our senses, and derived by them to the understanding, I call *SENSATION*.

And regarding reflection, Locke, acknowledged that one's soul or internal senses temper one's perceptions (Locke, 1996, p. 34):

> [T]he other fountain, from which experience furnishes the understanding with *ideas* is the *perception of the operations of our own minds* within us, as it is employed about the *ideas* it has got; which operations, when the soul comes to reflect on, and consider, do furnish the understanding with another set of *ideas*, which could not be had from things without: and such are, *perception, thinking, doubting, believing, reasoning, knowing, willing*, and all the different actings of our own minds; which we being conscious of, and observing in ourselves, do from these receive into our understandings, as distinct *ideas*, as we do from bodies affecting our senses. This source of *ideas*, every man has wholly in himself: and though it be not sense, as having nothing to do with external objects; yet it is very like it, and might properly enough be called internal sense. But as I call the other *sensation*, so I call this *REFLECTION*, the *ideas* it affords being such only, as the mind gets by reflecting on its own operations within itself.

As clear as Locke was about sensation and reflection being the "fountains of knowledge" (1996, p. 33) from which ideas spring, he was equally clear that one's mind is not a blank slate. There is a precursor to this knowledge and that precursor is one's experiences (Locke, 1996, p. 33):

> Let us then suppose the mind to be, as we say, white paper, void of all characters, without any *ideas*; how comes it to be furnished? Whence comes it by that vast store, which the busy and boundless fancy of man has painted on it, with an almost endless variety? Whence has it all the materials of reason and knowledge? To this I answer, in one word, from *experience*; in that, all our knowledge is founded; and from that it ultimately derives itself. Our observation employed either, about *external sensible objects* [i.e., sensations], *or about the internal operations of our minds, perceived and reflected on by ourselves* [i.e., reflection], *is that, which supplies our understandings with*

all the materials of thinking. These two are the fountains of knowledge, from whence all the *ideas* we have, or can naturally have, do spring.

Building on the ideas of Locke, David Hume, in *An Enquiry Concerning Human Understanding*, published in English in 1748, referred to experiences in terms of impressions, feelings, and sensations. Under the subheading Of the Origins of Ideas in *An Enquiry*, Hume wrote (Hume, 2007, pp. 7–8):

> Every one will readily allow, that there is a considerable difference between the perceptions of the mind ... and when he afterwards recalls to his memory this sensation, or anticipates it by his imagination. These faculties may mimic or copy the perceptions of the senses; but they never can entirely reach the force and vivacity of the original sentiment ... Here therefore we may divide all the perceptions of the mind into two classes or species, which are distinguished by their different degrees of force and vivacity. The less forcible and lively are commonly denominated Thoughts or Ideas. The other species want a name ... Let us use a little freedom, and call them Impressions ... By the term impression, then, I mean all our more lively perceptions ... And impressions are distinguished from ideas, which are the less lively perceptions, of which we are conscious ... In short, all the materials of thinking are derived either from our outward or inward sentiment: the mixture and composition of these belongs alone to the mind and will. Or, to express myself in philosophical language, all our ideas or more feeble perceptions are copies of our impressions or more lively ones.

To justify or to prove his argument, he wrote (Hume, 2007, p. 8):

> When we analyze our thoughts or ideas, however, compounded or sublime, we always find that they resolve themselves into such simple ideas as were copied from a precedent feeling or sentiment [i.e., experience]. Even those ideas, which, at first view, seem the most wide of this origin, are found, upon a nearer scrutiny, to be derived from it ... we shall always find that every idea which we examine is copied from a similar impression.

Contemporary economists have entered into the discussion about sources of knowledge and their writings reflect a relationship between the generation of knowledge and various forms of innate and acquired experiences. For example, Schultz wrote (1975, p. 828):

> Our knowledge of a person's abilities consists of inferences drawn from his performance. An ability is thus perceived as the competence and efficiency with which particular acts are performed.

Machlup (1980, p. 179) argued that formal education is only one source of knowledge; knowledge is also gained experientially and at different rates by different individuals. Individuals can accrue knowledge from their day-to-day experiences which (Machlup, 1980, p. 179):[4]

> ... will normally induce reflection, interpretations, discoveries, and generalizations ...

Foray (2004) offered a different, yet very important, taxonomy of types or categories of knowledge. Specifically, he suggested that (2004, p. 14):

> Much knowledge is produced by invention, that is, it does not exist as such in nature and is "produced" by man. Other types of knowledge stem from discoveries, that is, the accurate recognition of something which already exists but which was concealed. Invention is the result of production; discovery the result of revealing.

In Foray's (2004, p. 14) discussion involving the challenges facing the reproduction of knowledge, he acknowledged the related contributions of Polanyi:[5]

> Polanyi (1966), who introduced us to the concept of tacit knowledge, points out an essential aspect of knowledge that makes its reproduction difficult. Tacit knowledge cannot be expressed outside the action of the person who has it. In general, we are not even aware of the fact that we have such knowledge, or else we simply disregard it.

And Antonelli and Fassio (2014, p. 16) noted:

> [K]nowledge is not homogeneous. Instead, it should be regarded as a highly differentiated bundle of knowledge items.

KNOWLEDGE AND PRODUCTIVITY CAPITAL

The correlation analysis that follows might be viewed as a modification of previously written equation (6.4), written here as:

$$IO = F (HC, EC, SC, UC) \tag{9.1}$$

where IO refers to the innovative (i.e., knowledge-based) outputs of patents received and scientific publications, HC to human capital, EC to experiential capital, SC to social capital, and UC to university capital.[6]

Patents received and in-print scientific publication at the time of the surveys were normalized by dividing each observation by the year of the NRC survey minus the year of the Phase II award. This normalization accounts for the fact that a firm could receive a patent or publish a scientific paper over an extended period of time, and that period of time is related to the year the firm received the Phase II award relative to the year of the NRC survey. Below we refer to these normalized measures as *per year* measures.

Tables 9.1 through 9.5 show the descriptive statistics and the correlations related to the metrics in equation (9.1) for each of the SBIR agencies. The number of projects to which the correlation analyses apply, by funding agency, are different from the number of projects previously reported in the tables and figures because the correlation analysis in these tables does not apply to those projects previously defined in Chapter 7 as failing.[7]

Table 9.1 *Descriptive statistics and correlations among patents received per year and scientific publications per year from DOD projects and productive capital, n = 1,137*

Variable	Mean	Standard deviation	Minimum	Maximum
Patents received per year	0.800	0.214	0	2.857
Scientific publications per year	0.296	0.550	0	6.000

Productive capital	Patents received per year	Scientific publications per year
Number of employees (HC)	−0.059**	−0.053*
Number of previous awards (EC)	0.049*	0.063**
Additional funding (SC)	0.051*	0.061**
University involvement (UC)	0.133***	0.160***

Note: *** significant at the .01-level, ** significant at the .05-level, * significant at the .10-level.

Table 9.2   *Descriptive statistics and correlations among patents received per year and scientific publications per year from NIH projects and productive capital, n = 661*

Variable	Mean	Standard deviation	Minimum	Maximum
Patents received per year	0.195	0.455	0	4.50
Scientific publications per year	0.586	1.939	0	33.75

Productive capital	Patents received per year	Scientific publications per year
Number of employees (HC)	−0.013	0.018
Number of previous awards (EC)	−0.030	0.036
Additional funding (SC)	0.071*	0.056
University involvement (UC)	0.018	0.099***

Note: *** significant at the .01-level, * significant at the .10-level.

Table 9.3   *Descriptive statistics and correlations among patents received per year and scientific publications per year from NASA projects and productive capital, n = 235*

Variable	Mean	Standard deviation	Minimum	Maximum
Patents received per year	0.052	0.129	0	1.40
Scientific publications per year	0.307	0.467	0	2.727

Productive capital	Patents received per year	Scientific publications per year
Number of employees (HC)	0.048	-0.091
Number of previous awards (EC)	0.058	0.148**
Additional funding (SC)	−0.020	0.097
University involvement (UC)	−0.001	0.158***

Note: *** significant at the .01-level, ** significant at the .05-level.

Table 9.4   *Descriptive statistics and correlations among patents received per year and scientific publications per year from DOE projects and productive capital, n = 249*

Variable	Mean	Standard deviation	Minimum	Maximum
Patents received per year	0.122	0.248	0	2.50
Scientific publications per year	0.623	3.512	0	53.846

Productive capital	Patents received per year	Scientific publications per year
Number of employees (HC)	0.146**	0.089
Number of previous awards (EC)	0.098	0.494***
Additional funding (SC)	−0.012	−0.038
University involvement (UC)	0.156***	0.118*

Note: *** significant at the .01-level, ** significant at the .05-level, * significant at the .10-level.

Table 9.5   *Descriptive statistics and correlations among patents received per year and scientific publications per year from NSF projects and productive capital, n = 265*

Variable	Mean	Standard deviation	Minimum	Maximum
Patents received per year	0.168	0.279	0	2.50
Scientific publications per year	0.387	0.729	0	6.833

Productive capital	Patents received per year	Scientific publications per year
Number of employees (HC)	−0.026	−0.038
Number of previous awards (EC)	0.119**	0.176***
Additional funding (SC)	0.146***	−0.048
University involvement (UC)	0.047	0.110*

Note: *** significant at the .01-level, ** significant at the .05-level, * significant at the .10-level.

An inspection of the correlation coefficients across funding agencies shows that the statistical relationship between productivity capital and the defined knowledge outputs, IO, from equation (9.1) varies by funding agency. Among the DOD projects, human capital is negatively related to patents received and scientific publications; but experiential capital, social capital, and university capital are positively related to both of these knowledge outputs. If one views our measure of human capital as simply a measure of firm size, then the negative correlations suggest that the production of patents and scientific publications (i.e., the knowledge outputs) is being considered as more of a strategic element of behavior among smaller-sized firms than among larger-sized firms.[8] The other productive capital elements perhaps might be considered to be necessary aspects of firm behavior that are associated with internal and external experiences for the generation of the knowledge outputs being considered.

Among the NIH projects, the relationship between the elements of productive capital and knowledge outputs is generally weak, although for patents received the relationship is statistically stronger for social capital, and for scientific publications the relationship is statistically stronger for university capital.

Among the NASA projects, experiential capital and university capital are significantly related to scientific publications, and none of the other correlations are significant.

Among the DOE projects, university capital is significantly related to patents received and to scientific publications. In addition, human capital is significantly related to patents received, and experiential capital is significantly related to scientific publications.

Finally, among the NSF projects, experiential capital is significantly related to patents received and to scientific publications. The relationship is also significant between social capital and patents received, as well as between university capital and scientific publications.

The conclusion that we draw from the correlation analysis in this chapter is that no clear pattern emerges and more information is needed to understand, from both an economics perspective as well as a policy perspective, the determinants of patents received and scientific publications. Given that these knowledge outputs potentially have social benefits, additional study of this topic is clearly warranted.

SUMMARY AND WHAT FOLLOWS

An aspect of the social benefits associated with SBIR-funded Phase II projects is the technology-related knowledge that is transferred to society from the funded firms. Two mechanisms through which knowledge transfers occur are patents and scientific publications. The productive capital within a firm is related to the volume of such activity.

In Chapter 10, we describe the financial stakeholders to the technologies developed from the Phase II research projects. As discussed in Chapter 2, the funded firms are expected to pursue third-party funding to fund the commercialization of the technology resulting from the Phase II research project. To the extent that such third-party funding comes from the private sector prior to the Phase II award, one might conclude that the likelihood that the commercialized technology will be utilized is positive, and thus the commercialized technology will be a fourth measure of the social benefits of SBIR awards.

NOTES

1. While we will discuss in Chapter 14 the social benefits associated with the SBIR program, we note here that both of these knowledge outputs represent vehicles through which publicly funded research can be transferred into the public domain. Society benefits from the transfer of such knowledge.
2. The European Commission (2020) defined a number of knowledge transfer metrics, and scientific publications are the top-listed knowledge output. The European Commission refers to patents as a knowledge transfer channel.
3. The ideas expressed in this section draw directly from Audretsch and Link (2019a, 2019b).
4. Machlup (1962, pp. 21–2) proposed that a man has five types of knowledge: "(1) Practical knowledge: useful in his work, his decisions, and actions; (2) Intellectual knowledge: satisfying his intellectual curiosity; (3) Small-talk and pastime knowledge: satisfying the nonintellectual curiosity of his desire for light entertainment and emotional stimulation; (4) Spiritual knowledge: related to his religious knowledge of God and the ways to the salvation of the soul; and (5) Unwanted knowledge: outside his interests, usually accidentally acquired, aimlessly retained."
5. Machlup (1980) also discussed Polanyi's concept of tacit knowledge. Hess and Ostrom (2007, p. 8) have argued, citing Polanyi: "Acquiring and discovering knowledge is both a social process and a deeply personal process" (Polanyi, 1974, original 1958). See also Polanyi (1966) for further discussion of tacit knowledge.
6. Recall that data were not available to us for including in the NRC second-round databases on technical capital (TC).

7. Patents received (as opposed to patent applications) and in-print scientific publications (as opposed to research papers being submitted for publication) are the measures of IO in Tables 9.1 through 9.5. The survey data in certain NRC databases report, for some but not all funding agencies, patent applications and scientific papers submitted for publication. For consistency, patents received and in-print scientific publications are used in this chapter because they relate to all of the funding agencies. See Audretsch et al. (2019).

8. See Hayter and Link (2018) for a discussion about why small firms publish their innovative ideas.

10. Financial stakeholders in SBIR projects

INTRODUCTION

According to the Cambridge Dictionary, a *stakeholder* is "a person or group of people who own a share in a business."[1] And according to Merriam-Webster, a *stakeholder* is "one who is involved in or affected by a course of action."[2] We thus suggest, with regard to the scope and emphasis of this book, that the term *financial stakeholder* in relation to an SBIR project is a person or group of people who have a monetary investment in the research activity of a business/firm and thus who will be affected by the success of the firm's research activity.

The correlation analysis in Chapter 7 was interpreted to mean, at least among DOD research projects and NSF research projects, that those firms defined to embody social capital—as measured in terms of whether or not the firm received from private-sector entities prior to its Phase II award any R&D funding related to the technology being researched as a Phase II project—are in a statistical sense less likely to have their Phase II SBIR projects fail (see Tables 7.2 and 7.6). And among NASA research projects and DOE research projects, those firms defined to embody social capital are in a numerical sense rather than a statistical sense also less likely to have their Phase II SBIR research projects fail (see Tables 7.4 and 7.5). In this chapter, we consider the relationship between private-sector investments in Phase II research projects and—keeping in mind that the current statement of the purpose of the SBIR program is to increase private-sector commercialization of innovations derived from the Phase II SBIR award—whether or not the firm commercialized a technology from its Phase II research project.

DEFINING COMMERCIALIZATION

The survey instruments from the first and second rounds of NRC surveys all contained the same question about commercialization as defined in terms of the firm successfully bringing the Phase II technology to market. The survey question is:

> Has the company and/or licensee had any actual sales of products, processes, services, or other sales incorporating the technology developed during this project?

This survey question is only relevant to those Phase II projects that were not defined as failing. We define a binary variable for commercialization to equal 1 if there had been either sales of product(s), sales of processes(es), or sales of service(s) of the Phase II developed technology, and 0 otherwise.[3]

Table 10.1 shows descriptive statistics for the commercialization of a technology from a Phase II project, by funding agency. The mean value of the commercialization variable is the (estimated) probability that the new technology from a Phase II research project that did not fail was commercialized. The probability of commercialization ranged across the funding agencies from a low of 36.5 percent to a high of 64.5 percent.

Table 10.1   Descriptive statistics on the commercialization from a Phase II project, by funding agency

Funding agency	Mean	Standard deviation	Minimum	Maximum
DOD (n = 1,137)	0.566	0.496	0	1
NIH (n = 661)	0.419	0.494	0	1
NASA (n = 235)	0.613	0.488	0	1
DOE (n = 249)	0.365	0.483	0	1
NSF (n = 265)	0.645	0.479	0	1

STAKEHOLDERS AND COMMERCIALIZATION

As noted above, our measure of the presence of a stakeholder(s) is whether or not the Phase II-funded firm received from private-sector entitles prior to its Phase II award any R&D funding related to the technology being researched as a Phase II project. Previously, in Chapter 6, we used this metric to define social capital (SC). The question to ask is: Are projects with stakeholders more or less likely to commercialize? The presence of a stakeholder(s) means that there is an additional so-called set of eyes monitoring the due diligence of the firm in the conduct of its Phase II research. If this additional monitoring is effective, such projects that did not fail should be more likely to be successful in the market, and hence, more likely to commercialize.

Tables 10.2 through 10.6 show the correlation coefficients between our measure of commercialization and the presence of stakeholders. Also shown in these tables is the complementary mean value of the stakeholder metric for those projects that did commercialize and those that did not.

Table 10.2   *Relationship between stakeholders and the commercialization of DOD-funded Phase II projects, n = 1,137*

	Stakeholders	Commercialization
Stakeholders	1	
Commercialization	0.078***	1

Note: *** significant at the .01-level.

Variable	Mean	Standard deviation	Minimum	Maximum
Stakeholders when commercialization = 0 (n = 494)	0.136	0.343	0	1
Stakeholders when commercialization = 1 (n = 643)	0.194	0.396	0	1

Note: 1,137 of the 1,463 Phase II projects did not fail. See Table 7.1.

Table 10.3   *Relationship between stakeholders and the commercialization of NIH-funded Phase II projects, n = 661*

	Stakeholders	Commercialization
Stakeholders	1	
Commercialization	0.070*	1

Note: * significant at the .10-level.

Variable	Mean	Standard deviation	Minimum	Maximum
Stakeholders when commercialization = 0 (n = 384)	0.099	0.299	0	1
Stakeholders when commercialization = 1 (n = 277)	0.144	0.352	0	1

Note: 661 of the 820 Phase II projects did not fail. See Table 7.1.

Table 10.4   *Relationship between stakeholders and the commercialization of NASA-funded Phase II projects, n = 235*

	Stakeholders	Commercialization
Stakeholders	1	
commercialization	0.086	1

Variable	Mean	Standard deviation	Minimum	Maximum
Stakeholders when commercialization = 0 (n = 91)	0.088	0.285	0	1
Stakeholders when commercialization = 1 (n = 144)	0.146	0.354	0	1

Note: 235 of the 306 Phase II projects did not fail. See Table 7.1.

Table 10.5   *Relationship between stakeholders and the commercialization of DOE-funded Phase II projects, n = 249*

	Stakeholders	Commercialization
Stakeholders	1	
Commercialization	0.070	1

Variable	Mean	Standard deviation	Minimum	Maximum
Stakeholders when Commercialization = 0 (n = 158)	0.070	0.255	0	1
Stakeholders when Commercialization = 1 (n = 91)	0.110	0.314	0	1

Note: 249 of the 315 Phase II projects did not fail. See Table 7.1.

Table 10.6   *Relationship between stakeholders and the commercialization of NSF-funded Phase II projects, n = 265*

	Stakeholders	Commercialization
Stakeholders	1	
Commercialization	0.087	1

Variable	Mean	Standard deviation	Minimum	Maximum
Stakeholders when commercialization = 0 (n = 94)	0.245	0.432	0	1
Stakeholders when commercialization = 1 (n = 171)	0.327	0.471	0	1

Note: 265 of the 316 Phase II projects did not fail. See Table 7.1.

The correlations between stakeholders and commercialization are positive, as predicted, for all of the funding agency samples, but it is only statistically significant for DOD-funded and NIH-funded agency projects. Placing statistical inferences aside for the moment, one can also see from the tables that for every agency the mean value of stakeholders for the subsample of projects that did commercialize (= 1) is greater than for the subsample of projects that did not commercialize (= 0).

Stakeholders matter. Those Phase II projects for which the firm has stakeholders are more successful in terms of the funding agency fulfilling its Phase II mission to commercialize. And as we have posited, commercialization is a form of technology transfer and thus it represents the fourth measure of the social benefits from Phase II SBIR awards.

SUMMARY AND WHAT FOLLOWS

In this chapter the social benefits associated with SBIR-funded projects are technology transfers in the form of the commercialization of the developed technology. We showed that the likelihood of commercialization is greater when there are private-sector financial stakeholders (i.e., social capital).

In Chapter 11 we describe the distribution of technology sales among various segments of the market.

NOTES

1. See https://dictionary.cambridge.org/us/dictionary/english/stakeholder. Accessed March 26, 2022.
2. See https://www.merriam-webster.com/dictionary/stakeholder. Accessed April 3, 2022.
3. We discuss the markets for the sale of Phase II technologies in Chapter 11.

11. Market for SBIR developed technologies

In Chapter 10, we defined those projects for which the firm had realized sales of product(s), sales of processes(es), or sales of service(s) as being projects with a commercialized technology. However, in that chapter, there was no mention of the sector(s) to which the sales were made. In this chapter, we describe the sector in which the customers (i.e., buyers of products, processes, or services that were developed with SBIR support) of the Phase II SBIR-developed technologies operated.

The researchers who have studied the probability that a Phase II project will lead to a commercialized technology—see Table 4.2—have not provided information about the customer base for that technology. This is a relevant omission if one is interested in understanding the extent of direct (to final consumers) versus the indirect (to intermediate as discussed below) transfer of knowledge and thus of the breadth of social benefits associated with the SBIR program. While we do not offer insight into the social benefits of direct versus indirect knowledge transfer in this chapter, our descriptive findings emphasize the extent to which such dichotomously measured commercialization behavior exists.

The survey data allow us to define, by agency, the customer as being in the private sector, in the federal government sector (i.e., in a public-sector agency including the agency that funded the Phase II research), in state or local government, and/or in a foreign country (i.e., the technology was exported). The manner in which the survey question was written allowed a firm to report the portfolio of sectors of customers. For example, the survey question was:

> Approximately what percent of total sales from the technology developed during this project have gone to the following customers ...?

Tables 11.1 through 11.5 show, for project technologies that were commercialized and for which there was a reported distribution of sales, the mean percentage of sales to each of the four identified sectors. The percentages in these tables are based on the number of firms that responded to the survey question. A non-response to the question from a project respondent for whom the project was commercialized was interpreted to mean that the respondent did not know about the sector in which sales were made. Therefore, the percentages in these tables should be interpreted as referencing Phase II projects with sales to a particular sector as known by the respondent.[1]

Regardless of the respondent's reason for not responding to the survey question, it is clear that not all sales represented a direct transfer of knowledge, that is, not all sales went to customers who are in the private sector. A relatively large percentage of sales of DOD and NASA Phase II developed technologies might be described as indirect knowledge transfers, that is, those sales were to divisions within their own agency and across these two agencies. In addition, over 20 percent of NIH and DOE Phase II developed technologies were exported, and, with the caveat that the sample sizes are small, this pattern might best be described as an indirect domestic knowledge transfer or at worst as a technology transfer that is not supportive of domestic growth. One clear pattern from the percentages in the tables is that state and local governments represent the customer group with the smallest percentage of sales.

Table 11.1   Percent of sales of products, processes, and services from DOD-funded Phase II technologies, by sector, n = 625

Sector	Mean percent (%)
Private sector	26.3
Federal government	63.0
State and local government	1.0
Export	7.8

Note: 643 of the 1,137 Phase II projects that were not classified as failing commercialized from their research project (see Table 10.2); 625 projects reported information about their distribution of sales.

Table 11.2   *Percent of sales of products, processes, and services from NIH-funded Phase II technologies, by sector, n = 29*

Sector	Mean percent (%)
Private sector	60.8
Federal government	16.0
State and local government	5.2
Export	21.5

Note: 277 of the 661 Phase II projects that were not classified as failing commercialized from their research project (see Table 10.3); 29 projects reported information about their distribution of sales.

Table 11.3   *Percent of sales of products, processes, and services from NASA-funded Phase II technologies, by sector, n = 140*

Sector	Mean percent (%)
Private sector	36.4
Federal government	49.9
State and local government	0.59
Export	9.8

Note: 144 of the 235 Phase II projects that were not classified as failing commercialized from their research project (see Table 10.4); 140 projects reported information about their distribution of sales.

Table 11.4   *Percent of sales of products, processes, and services from DOE-funded Phase II technologies, by sector, n = 7*

Sector	Mean percent (%)
Private sector	57.1
Federal government	14.3
State and local government	0
Export	28.6

Note: 91 of the 249 Phase II projects that were not classified as failing commercialized from their research project (see Table 10.5); 7 projects reported information about their distribution of sales.

Table 11.5   *Percent of sales of products, processes, and services from NSF-funded Phase II technologies, by sector, n = 157*

Sector	Mean percent (%)
Private sector	56.6
Federal government	17.0
State and local government	2.7
Export	15.6

Note: 171 of the 265 Phase II projects that were not classified as failing commercialized from their research project (see Table 10.6); 157 projects reported information about their distribution of sales.

SUMMARY AND WHAT FOLLOWS

In Chapter 10, the commercialization of developed technologies was discussed in terms of the social benefits associated with the SBIR program. In this chapter, we describe the economic sectors in which the sales occur of products, processes, or services that were developed with SBIR support. The sectors differ across SBIR funding agencies, which suggests that the innovative outputs from SBIR-funded research have a broad and varied impact on society.

In Chapter 12, we introduce a normative dimension to our study of the SBIR program by addressing any unanticipated or unintended consequences associated with the program. The unanticipated consequences that we address relate to one of the social benefit activities previously discussed, namely, the patenting of publicly funded research technologies by private-sector firms.

NOTE

1.   In other words, the data were coded as 0 if the actual percentage of a project's sales were zero to that sector or if the actual percentage of a project's sales to that sector was not reported.

PART III

SBIR program: policy perspectives

12. Unanticipated consequences

INTRODUCTION

What do policy makers and legislators think about when crafting and adopting a new legislative initiative or modifying an existing legislative initiative? Answers to this question will likely vary among individuals, and the answers given will likely reflect one's experiences as, or working with, policy makers and legislators. In this chapter, we take the position that this is an important question to ask, especially when public resources are involved as they are by the agencies supporting the SBIR program and funding Phase I and Phase II research projects.

Regardless of the answers offered to this question, the answers will likely be based on one's own implicit set of assumptions. At one end of such an assumption set is the proposition that the objective function of policy makers and legislators is to act solely in a manner that maximizes their probability of being re-elected to public office. At the other end of the assumption set is the proposition that the objective function of policy makers and legislators is to act solely in a manner that maximizes the public or social good.

Countries have different legislative processes and institutions that support public officials. Of course, a policy maker or legislator might in fact believe that his/her being re-elected to public office de facto maximizes the public or social good. If such is the case, his/her orientation, be it myopic or not, might result in encouraging the allocation of public moneys toward projects that are able to demonstrate outcomes within a short period of time (i.e., a period of time before the next election or appointment), regardless of the overall social benefits that might be associated with projects that require more time to come to fruition.

Wherever most policy makers and legislators fall along this assumption set, there will likely be anticipated as well as unanticipated consequences associated with any if not all of their legislated actions.

The concept of unanticipated consequences of an action, be it policy related or not, has a long history in the literature. And this literature has a history that traces at least to the writings of Frédéric Bastiat.[1] Bastiat wrote in 1848 in *Selected Essays on Political Economy* (1995, p. 1):

> In the economic sphere an act, a habit, an institution, a law produces not only one effect, but a series of effects. Of these effects, the first alone is immediate; it appears simultaneously with its cause; *it is seen.* The other effects emerge only subsequently; *they are not seen;* we are fortunate if we *foresee* them.
>
> There is only one difference between a bad economist and a good one: the bad economist confines himself to the *visible* effect; the good economist takes into account both the effect that can be seen and those effects that must be *foreseen.*
>
> Yet this difference is tremendous; for it almost always happens that when the immediate consequence is favorable, the later consequences are disastrous, and vice versa. Whence it follows that the bad economist pursues a small present good that will be followed by a great evil to come, while the good economist pursues a great good to come, at the risk of a small present evil.

Bastiat went on to write that through experience (and as we discussed in Chapter 9, one's experiences are a source of one's knowledge) one learns or one eventually learns to foresee or anticipate consequences (1995, p. 2):

> Experience teaches efficaciously but brutally. It instructs us in all the effects of an act by making us feel them, and we cannot fail to learn eventually, from having been burned ourselves, that fire burns. I should prefer, in so far as possible, to replace this rude teacher with one more gentle: foresight.

Perhaps, then, based on Bastiat, in the world of legislative initiatives the more experienced policy makers or members of a legislative group have greater foresight than those with less experience. But, given that group dynamics are what they are, it is not always the more experienced policy maker or member of a legislative group whose voice is the loudest; and it is not always the case that experience is positively correlated with persuasiveness.

Merton (1936), an eminent sociologist, wrote in his famous essay "The Unanticipated Consequences of Purposive Social Action" the following (1936, p. 898):[2]

> The most obvious limitation to a correct anticipation of consequences of action is provided by the existing state of knowledge. The extent of this limitation may be best appreciated by assuming the simplest case where this lack of adequate knowledge is the sole barrier to a correct anticipation.

Merton's proposition can be restated as follows: the correct anticipation of the consequences associated with an action is driven by adequate knowledge. We believe that we have made the case in some of the previous chapters that knowledge follows from a variety of experiences. Because one's experiences are nevertheless limited, we suggest that anticipated as well as unanticipated consequences will inevitably be associated with policy makers' and legislators' actions regarding any technology policy, and the SBIR program is not likely to be an exception.

Some of the anticipated consequences associated with the funding of a Phase II research project have already been discussed in previous chapters, namely, the commercialization of a Phase II-funded technology. And the commercialization of a new technology is correctly anticipated to be a form of knowledge transfer through new technology, and as such it will inevitably leverage economic growth and thus have positive social consequences.

However, there could be unanticipated consequences associated with aspects of the SBIR program, and in particular with funded Phase II projects, and that is the topic of this chapter.

One might think of unanticipated consequences in terms of innovative outputs from Phase II projects for which the social rate of return is less than the social hurdle rate.[3]

Consider Figure 12.1 to be a framework to identify unanticipated consequences from an SBIR award. Four quadrants are shown in the figure: I, II, III, and IV. Also shown in the figure is the society's hurdle rate. Any project with an expected rate of return that is less than this social rate should not, from society's perspective, be pursued, that is, should not be supported with SBIR funds, because its use of resources is less than socially optimal. R^* represents the minimum level of R&D investment for the firm to meet its hurdle rate.

The literature is replete with two-by-two matrices to explain various phenomena. One might view Figure 12.1 as a two-by-two matrix where Quadrant I refers to R&D investments that the private sector is willing to undertake and that the public sector wants to see the private sector undertake. Quadrant III refers to R&D investments that the private sector does not want to pursue, and the public sector embraces the private sector for that decision. Quadrants II and IV are mixed. One might refer to R&D investments in Quadrant II as being policy worthy: the private sector will not undertake the R&D investments but the public sector would like to see the private sector do so. Thus, the public sector will adopt a policy to provide an incentive to the private sector for making the investment,

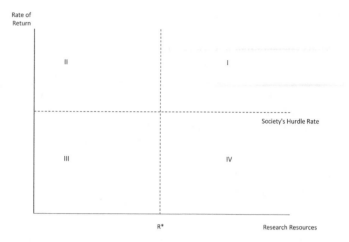

Figure 12.1   A framework to identify unanticipated consequences

thus moving it into Quadrant I. R&D investments in Quadrant IV are also policy worthy. In this case the private sector wants to undertake the investment but the public sector does not view such investment as being in society's best interest. The appropriate policy would be one that gave the private sector an incentive to invest elsewhere.[4]

If the firm has a research budget that is slightly less than R*, this firm will not pursue the research project that it is considering. If the firm has a research budget that is greater than R*, the firm would consider pursuing the project and it expects a private rate of return from the use of those recourses would exceed its private hurdle rate (see Figure 2.3). For illustration purposes, we assume that if resources available for the project are greater than R*, the expected private rate of return will be greater than the firm's private hurdle rate. Thus, a project in Quadrant I in Figure 12.1 has, or is expected to have, a rate of return that is greater than the firm's private hurdle rate and greater than society's hurdle rate. A project in Quadrant II only has a rate of return that is greater than society's hurdle rate. A project in Quadrant III has a rate of return less than the firm's private hurdle rate and less than society's hurdle rate. Finally, a project in Quadrant IV only has a rate of return that is greater than the firm's private hurdle rate.

Assume that the firm applies for a Phase I award and receives it, and assume that the firm successfully demonstrates proof of concept from a technical perspective. The firm then applies for a Phase II award and receives it. However, when the funded research project is completed, the developed technology is associated with a private rate of return that is above the firm's private hurdle rate but the social rate of return then is less than society's hurdle rate; the developed technology is in Quadrant IV rather than in Quadrant I. In this scenario, the SBIR program evaluated a potential project, funded the project through due diligence and thus allocated to the firm the research resources needed to entice the firm to undertake the project (give the firm financial resources that, with internal resources, push the resources available for the project to be greater than R*). But, in this scenario, the developed technology resulted in the use of social resources that turned out to be less than socially optimal (e.g., the firm properly developed the technology for a 3D printer that could be used to produce illegal weapons).

PATENTING AS AN EXAMPLE OF UNANTICIPATED CONSEQUENCES

The unanticipated consequence that we focus on in this chapter relates to the fact that a firm that received an SBIR Phase II research award is permitted to patent some or all elements of the new technology that resulted from the research funded by the award. In other words, the SBIR program permits firms to use public moneys to delimit the actual transfer, but not the potential transfer, of the knowledge embodied in a commercialized technology for personal gain.[5]

Concerns have long been expressed about universities patenting new technologies that result from research supported by public moneys. For example, some of the concerns expressed have identified the Bayh-Dole Act as culprit for allowing universities to capture monopoly rights through patents to publicly funded research results. To the best of our knowledge, the SBIR program, through its allowance of patenting by funded firms, has not *yet* been targeted.

Link et al. (2018, p. 34) wrote the following about the Bayh-Dole Act and about the universities supported through public research awards;

we suggest in this chapter that the same argument might apply to firms supported through the SBIR program:

> The role of Bayh-Dole in distorting the innovation system is apparent. Pursuit of intellectual property has generated conflicts between universities over precedence and restricted the use of knowledge by both university and industry researchers.

Figures 12.2 through 12.6 show by award year the number of patents received by firms, which have not been characterized as having a Phase II project fail, that are related to their funded Phase II research. The data that support the construction of these figures were discussed in Chapter 9.[6] The upper segment in each figure shows the number of such patents to firms that were able to commercialize their technology, and the lower segment in each figure shows the number of patents to firms that were not able to commercialize their technology (or at least did not commercialize their technology as of the time of the NRC surveys). Imposed on both the upper segment and the lower segment of each figure are linear trend lines. While these data and the trends in the data do not in themselves quantify a loss of social benefits from Phase II SBIR awards, they do at a minimum raise a number of questions that have yet to be considered to the best of our knowledge by either academics or policy makers.

First, across the five agencies being considered, there is suggestive evidence that firms are either increasing (NIH, DOE, and NSF) or at least not noticeably decreasing (e.g., DOD and NASA) their patenting activity. While consumers do have access to any Phase II developed technology that is sold in the marketplace, they do not have access without a licensing fee to the patented knowledge embodied in the developed technology. However, consumers do not have access to any not fully-developed technology because it is not commercialized.

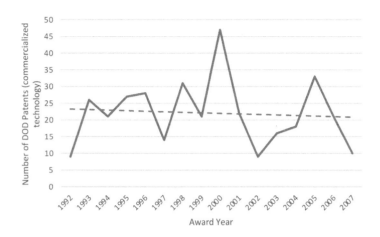

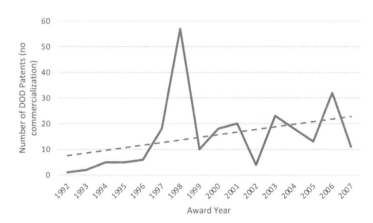

Note: The dashed lines are linear trend lines.

Figure 12.2   *Trends in patents received by DOD-funded firms based on Phase II research, by commercialization status, n = 1,137*

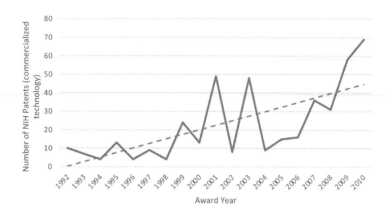

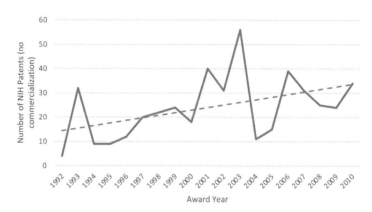

Note: The dashed lines are linear trend lines.

Figure 12.3   Trends in patents received by NIH-funded firms based on Phase II research, by commercialization status, n = 661

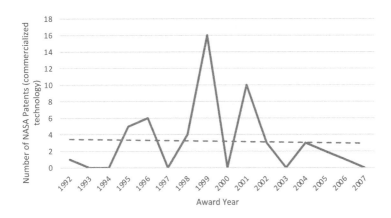

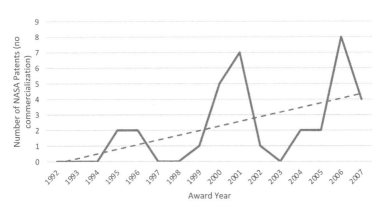

Note: The dashed lines are linear trend lines.

Figure 12.4 *Trends in patents received by NASA-funded firms based on Phase II research, by commercialization status, n = 235*

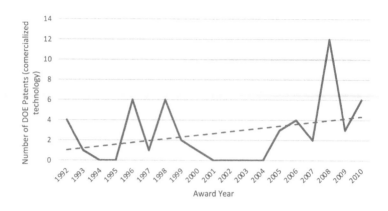

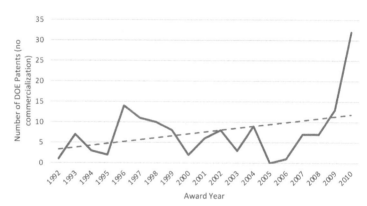

Note: The dashed lines are linear trend lines.

Figure 12.5   *Trends in patents received by DOE-funded firms based on Phase II research, by commercialization status, n = 249*

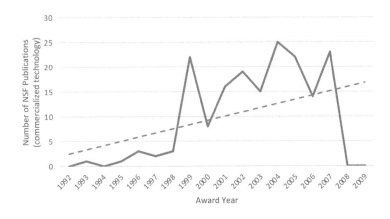

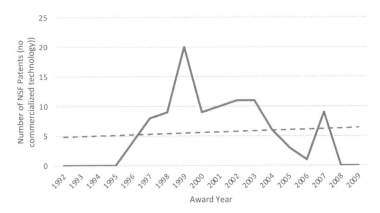

Note: The dashed lines are linear trend lines.

Figure 12.6   *Trends in patents received by NSF-funded firms based on Phase II research, by commercialization status, n = 265*

The information in the lower segment of each figure might suggest that there are unanticipated consequences of allowing SBIR-funded firms to patent their research. The data in the lower segment of each figure show the patenting activity of firms that were not able to commercialize their technology. For some agencies (e.g., DOE), the number of patents awarded to non-commercialized technologies is even greater than for commercialized technologies. To the extent that the underlying research knowledge in the non-commercializing firms has possible probative value to other firms, were that knowledge open sourced into the public domain, then one should consider the impact on the economy for that publicly funded knowledge being withheld.

Should consumers have access to one or both sets of SBIR-funded technology? On the one hand, we suggest that perhaps they should because both sets of technology were funded with public moneys. On the other hand, if third-party development resources were involved in bringing the Phase II developed technology to market, those investors are entitled to a return on their investments. The information from either the upper segment or lower segment of each figure is insufficient to give weight to one of these arguments over the other.

We are not suggesting that the patenting privilege be removed from firms that are or are not successful in developing new technologies or portions of new technologies. We do wonder, however, if those policy makers and legislatures who envisioned the SBIR program thought about open-source technology or about the public ownership of publicly funded technology given that the shelf-life of new technology might be short. At a minimum, the purpose of this chapter is to raise issues that perhaps should be considered.

SUMMARY AND WHAT FOLLOWS

We introduced a normative dimension to our study of the SBIR program and its impact on firm behavior. The aspect of firm behavior that was addressed was patenting. While patenting is a form of knowledge transfer for some, it might be viewed as a deterrent to knowledge transfer to others and thus allowing firms to patent their SBIR-funded technologies might be viewed as an unanticipated consequence of the program.

Chapter 13 focuses on some additional elements of the social benefits associated with the SBIR program. The approach of Chapter 13 is to adopt a counterfactual situation to determine whether or not a Phase II-funded firm would have proceeded to undertake the Phase II research project in the absence of SBIR funding. The counterfactual implications of a firm undertaking the Phase II project absent SBIR funding provides an additional glimpse into the social importance of the SBIR program.

NOTES

1. Unanticipated consequences associated with a legislated policy are not necessarily at odds with the concept of government failure (Wolf, 1988; Le Grand, 1991; Dolfsma, 2011). See Chapter 13.
2. Merton (1936, p. 894) also wrote: "In some one of its numerous forms, the problem of the unanticipated consequences of purposive action has been treated by virtually every substantial contributor to the long history of social thought. Some of the modern theorists, though their contributions are by no means of equal importance, are: Machiavelli, Vico, Adam Smith (and some later classical economists), Marx, Engels, Wundt, Pareto, Max Weber, Graham Wallas, Cooley, Sorokin, Gini, Chapin, von Schelting."
3. A hurdle rate corresponds to the rate of return below which a firm or society would not want a project to be undertaken because the return to the use of scarce resources is less than the return to the next best alternative use of the resources.
4. R&D investments that fall in Quadrant IV are a topic for another day. In that quadrant the public sector has to grapple with how flexible it can be on social hurdle rates.
5. We are aware of the argument that the revenues associated with the licensing of a Phase II developed technology might be used by the firm to further its own research and to bring more new technologies to market. At a minimum, we are of the opinion that it is not necessarily the case that the social benefits of the funded firm bringing new technologies to the market are greater than if the underlying aspects of Phase II developed technology could be open sourced, or if not, could be re-engineered and if successful other firms would be bringing new technologies to the market.
6. The underlying data are counts of patents. These patent data have not been weighted by citation counts in an effort to approximate the social value of the patents.

13. A counterfactual analysis

INTRODUCTION

Fundamental to public support of economic activity is the public sector's awareness of its accountability for its use of public resources. In our case in this book the public resources are the funding dollars to support Phase II awards from the SBIR programs. The concept of public accountability can be traced in the United States as far back as President Woodrow Wilson's reforms, and in particular to the Budget and Accounting Act of 1921 (Public Law 67–13).[1] This Act not only required the President of the United States to transmit to Congress a detailed budget on the first day of each regular session, but also it established what is known today as the Government Accountability Office (GAO) to settle and adjust all accounts of the government.[2]

The GAO has a long-standing tradition and a well-documented history of efforts to improve government agency management through performance measurement. In February 1985, the GAO issued a report entitled *Managing the Cost of Government—Building an Effective Financial Management Structure*, which emphasized the importance of systematically measuring performance as a key area to ensure a well-developed financial management structure (1985, p. 5):

Systematic Measurement of Performance
  Effective management of resources requires examining the results of government activities as well as their costs. An integrated and disciplined financial management system that provides consistent data on cost and performance is essential to help both Congress and the executive branch assess the efficiency and effectiveness of government operations.

A first step toward understanding program performance is to understand the distinction between program assessment and program evaluation. Although the terms *assessment* and *evaluation* are frequently used interchangeably, especially with reference to public-sector activities, many if not most economists believe that a clear distinction is warranted not only to ensure consistency of use of the terms but also to ground them both in concepts fundamental to economics (Link and Scott, 1998).

- Program assessment is based primarily on the criterion of effectiveness: Has the program met its stated goals and objectives; have its designated outputs been achieved?
- Program evaluation is based on the criterion of efficiency: How do the social benefits or outcomes associated with the program compare to the social costs?

A second step toward understanding program performance is to understand which program evaluation method is most appropriate for a study of the program in question. One method often used is what is known as a counterfactual evaluation method. The relevant evaluation question is: *Are public-sector investments a more efficient way of generating the technology than private-sector investments would have been?*

COUNTERFACTUAL INFORMATION ABOUT PHASE II PROJECTS

Table 13.1 shows the responses of firms with Phase II research projects that answered the counterfactual survey question:[3]

In the absence of the Phase II SBIR award, would your firm have undertaken research on this Phase II project?

Table 13.1   Percent of Phase II projects that would have been undertaken in the absence of SBIR funding

Funding Agency (n)	Percent of firms that responded "definitely yes" or "probably yes" to the statement: "In the absence of this Phase II SBIR award, would your firm have undertaken research on this Phase II project?"
DOD (1,137)	11.2%
NIH (661)	12.9%
NASA (235)	12.3%
DOE (249)	10.0%
NSF (265)	13.6%

The percentages in Table 13.1 show a double-digit affirmative response by firms funded by all of the funding agencies. For example, among the 1,137 Phase II projects funded by DOD, 11.2 percent reported that absent the SBIR funding they would have undertaken research on the Phase II project. The other percentages in the table are interpreted in the same way, and all of the percentages are approximately of the same level of magnitude.

Tables 13.2 through 13.6 describe separately selected characteristics about those firms that would have undertaken their Phase II research project absent SBIR funding and those firms that would not have undertaken their Phase II research project.

Table 13.2   *Characteristics of firms that would have (n = 127) / would not have (n = 1,010) undertaken the Phase II project in the absence of DOD SBIR funding*

Firm Characteristic	
Women-owned	
Would have undertaken the project absent DOD SBIR funding	11.0%
Would not have undertaken the project absent DOD SBIR funding	10.1%
Minority-owned	
Would have undertaken the project absent DOD SBIR funding	10.2%
Would not have undertaken the project absent DOD SBIR funding	9.8%
Number of employees at the time of the award (HC)	
Would have undertaken the project absent DOD SBIR funding	28.54
Would not have undertaken the project absent DOD SBIR funding	33.80
Number of previous Phase II awards related to the funded technology (EC)	
Would have undertaken the project absent DOD SBIR funding	1.39
Would not have undertaken the project absent DOD SBIR funding	1.31
Received private-sector funding related to the funded technology prior to the Phase II award (SC)	
Would have undertaken the project absent DOD SBIR funding	26.8%
Would not have undertaken the project absent DOD SBIR funding	15.6%

Table 13.3   *Characteristics of firms that would have (n = 85) / would not have (n = 576) undertaken the Phase II project in the absence of NIH SBIR funding*

Firm Characteristic	
Women-owned	
Would have undertaken the project absent NIH SBIR funding	2.4%
Would not have undertaken the project absent NIH SBIR funding	14.9%
Minority-owned	
Would have undertaken the project absent NIH SBIR funding	5.9%
Would not have undertaken the project absent NIH SBIR funding	5.6%
Number of employees at the time of the award (HC)	
Would have undertaken the project absent NIH SBIR funding	25.65
Would not have undertaken the project absent NIH SBIR funding	19.43
Number of previous Phase II awards related to the funded technology (EC)	
Would have undertaken the project absent NIH SBIR funding	0.82
Would not have undertaken the project absent NIH SBIR funding	1.44
Received private-sector funding related to the funded technology prior to the Phase II award (SC)	
Would have undertaken the project absent NIH SBIR funding	17.6%
Would not have undertaken the project absent NIH SBIR funding	10.9%

Table 13.4   *Characteristics of firms that would have (n = 29) / would not have (n = 206) undertaken the Phase II project in the absence of NASA SBIR funding*

Firm Characteristic	
Women-owned	
Would have undertaken the project absent NASA SBIR funding	10.3%
Would not have undertaken the project absent NASA SBIR funding	8.7%
Minority-owned	
Would have undertaken the project absent NASA SBIR funding	6.9%
Would not have undertaken the project absent NASA SBIR funding	6.8%
Number of employees at the time of the award (HC)	
Would have undertaken the project absent NASA SBIR funding	39.14
Would not have undertaken the project absent NASA SBIR funding	38.49
Number of previous Phase II awards related to the funded technology (EC)	
Would have undertaken the project absent NASA SBIR funding	1.90
Would not have undertaken the project absent NASA SBIR funding	1.17
Received private-sector funding related to the funded technology prior to the Phase II award (SC)	
Would have undertaken the project absent NASA SBIR funding	17.2%
Would not have undertaken the project absent NASA SBIR funding	11.7%

Table 13.5   *Characteristics of firms that would have (n = 25) / would not have (n = 224) undertaken the Phase II project in the absence of DOE SBIR funding*

Firm Characteristic	
Women-owned	
Would have undertaken the project absent DOE SBIR funding	8.0%
Would not have undertaken the project absent DOE SBIR funding	5.4%
Minority-owned	
Would have undertaken the project absent DOE SBIR funding	8.0%
Would not have undertaken the project absent DOE SBIR funding	8.5%
Number of employees at the time of the award (HC)	
Would have undertaken the project absent DOE SBIR funding	40.56
Would not have undertaken the project absent DOE SBIR funding	28.19
Number of previous Phase II awards related to the funded technology (EC)	
Would have undertaken the project absent DOE SBIR funding	2.04
Would not have undertaken the project absent DOE SBIR funding	1.16
Received private-sector funding related to the funded technology prior to the Phase II award (SC)	
Would have undertaken the project absent DOE SBIR funding	8.0%
Would not have undertaken the project absent DOE SBIR funding	8.5%

Table 13.6   *Characteristics of firms that would have (n = 36) / would not have (n = 229) undertaken the Phase II project in the absence of NSF SBIR funding*

Firm Characteristic	
Women-owned	
Would have undertaken the project absent NSF SBIR funding	11.1%
Would not have undertaken the project absent NSF SBIR funding	10.0%
Minority-owned	
Would have undertaken the project absent NSF SBIR funding	19.4%
Would not have undertaken the project absent NSF SBIR funding	10.5%
Number of employees at the time of the award (HC)	
Would have undertaken the project absent NSF SBIR funding	31.31
Would not have undertaken the project absent NSF SBIR funding	14.47
Number of previous Phase II awards related to the funded technology (EC)	
Would have undertaken the project absent NSF SBIR funding	1.42
Would not have undertaken the project absent NSF SBIR funding	1.24
Received private-sector funding related to the funded technology prior to the Phase II award (SC)	
Would have undertaken the project absent NSF SBIR funding	41.7%
Would not have undertaken the project absent NSF SBIR funding	27.9%

Looking across the funding agency-specific tables in a non-statistical manner, a larger percentage of firms that would have proceeded with their Phase II research absent SBIR funding were owned by women. The only exception was among those Phase II projects funded by NIH. Again, looking across the funding agency-specific tables, a larger percentage of firms that would have proceeded with their Phase II research absent SBIR funding were minority-owned. The only exception was among those Phase II projects funded by the DOE.

With the exception of DOD-funded projects, firms that reported that they would proceed with their Phase II research project in the absence of SBIR funding had a greater human capital (HC) endowment, measured in terms of the number of employees at the time of the Phase II award.

With the exception of NIH-funded projects, firms that would have proceeded with their Phase II research project in the absence of SBIR funding had previously received more Phase II award related to the funded technology, or equivalently, had a greater endowment of experiential capital (EC).

With the exception of DOE-funded projects, among firms that were more likely to proceed with their Phase II research project in the absence of SBIR funding, a larger percentage had an endowment of social capital (SC), measured in terms of having received private-sector funding related to the technology being researched.

One might conclude from the above descriptive findings that there was a government failure in the sense that across all agencies, a number of Phase II research projects was unnecessarily funded.

One might think of government failure as purposeful public-sector involvement in a market economy that results in the market economy being less efficient than before the public sector's involvement.[4] Dolfsma (2011, p. 597) wrote:

> The non-exhaustive list of four different ways in which government can fail will draw mostly on scholarly work in the philosophy of law. When formulating rules, then, government can be (1) too specific, (2) too broad, (3) arbitrary, or (4) setting out rules that conflict with other rules it has set out to address other, related (possibly primarily non-economic) issues possibly for the same practice.

Dolfsma's view, which draws on the literature on government failure—Wolf's (1988) writings in particular—is that government failure results from ill-formed rules (i.e., ill-formed policies that set forth guidelines).

Firms responding to the counterfactual question posed on the surveys reported that their SBIR award was not a necessary condition for pursing their research projects. Perhaps so, however, as shown in Table 13.7, the firms that reported that they would have gone it alone, so to speak, also reported that the completion of their Phase II research would have been delayed by nearly a year or more. In fact, as reported in the table, the range of mean responses to the length of the delay in completing their Phase II research ranged from 11 months to nearly 17 months.

Table 13.7   Implications of undertaking the Phase II project in the absence of SBIR funding

Funding Agency	The number of months that the Phase II project would have been delayed in the absence of SBIR funding as reported by firms that would have undertaken the project in the absence of SBIR funding
DOD (125/127)	11.0
NIH (82/85)	12.0
NASA (29/29)	16.8
DOE (25/25)	16.7
NSF (36/36)	15.7

Note: Number of observations in parentheses is the number of firms that responded to this survey question and the number of firms that would have undertaken the project in the absence of the SBIR funding.

And, with the exception of DOE-funded projects, a larger percentage of Phase II projects were successful, meaning that the resulting technology was commercialized, by firms that reported that they would have under-taken their Phase II research project in the absence of SBIR fundings. See Table 13.8.

Table 13.8   *Commercialization success of undertaking and not undertaking the Phase II project in the absence of SBIR funding*

Counterfactual situation	DOD	NIH	NASA	DOE	NSF
Would have undertaken the project absent SBIR funding	59.1%	48.2%	69.0%	24.0%	69.4%
Would not have undertaken the project absent SBIR funding	56.2%	41.0%	60.2%	37.9%	63.8%

Note: Number of observations for each counterfactual situation by agency are reported in Tables 13.2 through 13.7.

Taking the above descriptive findings as a whole, it appears that a tradeoff might exist between allocating publicly funded Phase II awards to firms that would have undertaken the Phase II research in the absence of SBIR fundings and bringing a new technology to market sooner than would have otherwise been the case. Stated differently, perhaps a correct interpretation of our descriptive findings is that for some firms, the allocation of Phase II research awards is a means to bring new technology to the market sooner rather than later.

SUMMARY AND WHAT FOLLOWS

In this chapter, the likelihood that a SBIR-funded research project would have been undertaken by the firm in the absence of the Phase II award, and the characteristics of firms doing so was described.

In Chapter 14, we summarize our description of social benefits associated with the SBIR program, and we suggest a roadmap for future research about the SBIR program.

NOTES

1.   U.S. President Warren G. Harding signed the Budget and Accounting Act of 1921 on June 10 of that year. This Act created the Bureau of the Budget, known today as the Office of Management and Budget (OMB): "There is hereby created in the Treasury Department a Bureau to be known as the Bureau of the Budget ... the bureau when duly authorized, shall ... have access to, and the right to examine, any books, documents, papers, or records of any such department or agency." However, one might argue that the initial concept of fiscal accountability in the United States is rooted in the fundamental principle of representation "of the people, by the people." President Abraham Lincoln stated in his Gettysburg Address on November

19, 1863: "This nation, under God, shall have a new birth of freedom—and that government of the people, by the people, and for the people, shall not perish from the earth."

2.   In 1921, the Act created what is known today as the Government Accountability Office (GAO): "There is created an establishment of the Government to be known as the General Accounting Office, which shall be independent of the executive departments ... All claims and demands whatever by the Government of the United States or against it, and all accounts whatever in which the Government of the United States is concerned ..." are the responsibility of the Office.

3.   Affirmative responses to this question (= 1) definitely would have undertaken the Phase II project or probably would have undertaken the project. Non-affirmative responses (= 0) definitely would not have undertaken the Phase II project, probably would not have undertaken the Phase II projects, or uncertain about undertaking the Phase II project.

4.   An accessible comparison of government failure to market failure is in Winston (2006).

14. Concluding remarks

In this book we have described the SBIR program from three distinct perspectives: the institutional perspective, the empirical perspective, and the policy perspective. The institutional perspective focused on the legislative history of the program and how the program fits into the larger framework of U.S. technology policy. Our empirical perspective was built on the data collection efforts requested by Congress and undertaken by the NRC. We uncovered interesting and previously unknown statistical patterns among firm characteristics, project characteristics, and the different types of productive capital employed by firms conducting their SBIR-funded research projects. The third and final perspective was policy. We discussed potential unintended consequences of the program and provided a first step toward a counterfactual policy analysis. In the current chapter, we offer some concluding remarks and suggest a roadmap for future research.

One overarching conclusion from the empirical analyses is that it is generally difficult, if not impossible, to identify statistical patterns and relationships that are consistent across *all* five SBIR funding agencies contained in the NRC databases (DOD, NIH, NASA, DOE, and NSF). We encountered this, for example, when analyzing the ownership status of SBIR-funded firms in Chapter 5, and the correlations between human capital, experiential capital, and social capital in Chapter 6. One of the more consistent findings was that project failure, as defined in Chapter 7, has declined over time for SBIR-funded firms across all agencies. Larger firms, as measured by the number of employees, might be at increased risk of project failure, whereas experience, as measured by previous SBIR awards, appears to shield against such risk.

Given the substantial, unexplained heterogeneity across funding agencies, it is clear that a better understanding of the determinants of heterogeneity is needed. We suggest this as an important topic for future study, as it would help clarify which agencies are more or less successful in translating SBIR funding into social benefits, and why. To make progress in this area, it will likely be necessary to have more detailed information about the specific nature of the SBIR-funded research projects and the involved technologies, and the associated risks and rewards. This, in turn, may require new and more focused data collection efforts.

To conduct a comprehensive assessment of the SBIR program, it is important to consider the program's distinct social benefits. One such benefit has been the transfer of SBIR-type policies to other countries. As discussed in Chapter 2, the origins of the SBIR program can be traced to the Small Business Innovation Act of 1982, but it was not until 1993 that a similar program was adopted in South Africa. Since then, eight additional countries have implemented technology and innovation policies that are inspired and informed by the SBIR program. While international policy diffusion was not an original policy goal, such diffusion can nonetheless be considered a social benefit that extends beyond national borders. Possible directions for future research in this area include a characterization and comparison of SBIR-type programs across countries, and an empirical evaluation of their implementation successes (or failures).

Other social benefits of the SBIR program are the facilitation of knowledge transfer to society through patenting and scientific publications. A consistent finding across agencies, as documented in Chapter 9, is that the involvement of university collaborators or resources in the SBIR Phase II-funded research is associated with a larger number of scientific publications. This is hardly surprising, given the publication incentives that exist in academic environments.

Ultimately, though, our analysis cannot determine whether the SBIR program brought the level of knowledge transfer closer to, or farther from, a socially optimal level. Put differently: Did the program result in "too much" patenting, thereby stifling innovation, or did it encourage the creation of knowledge that would not have seen the light of day otherwise? And should we regard patents resulting from SBIR-funded research as an unintended consequence of the program? Chapter 12 highlighted that the debate about the pros and cons of intellectual property protection of publicly supported R&D is ongoing and remains to be settled. As such, trying to address these questions empirically and in the context of the SBIR program is an important topic that can generate substantial future research.

A fourth and final social benefit of the SBIR program that we discuss in this book is the transfer of technology through commercialization of the SBIR-funded research, where we defined commercialization in Chapter 10 as the sale of products, processes, or services that incorporate the technology developed during the SBIR Phase II project. A consistent finding across funding agencies is that the presence of private-sector funding received for related technologies prior to the SBIR Phase II award, or social capital, is associated with an increased likelihood of commercialization. This suggests that private stakeholders, perhaps through increased monitoring or due diligence, may facilitate technology transfer to the market by increasing the likelihood of commercialization.

Here, too, further research will be beneficial. For example, additional insights into the causal determinants of commercialization success, as well as consideration of a wider range of program outcome metrics, will help policy makers and administrators increase the overall effectiveness and social benefits of the SBIR program. There are various methods for researchers to adopt for pursuing these research ideas. Case studies, surveys, and access to related public-domain information take time but we believe that there are low-hanging fruits to be harvested.

We have not discussed Phase I projects in any detail in this book, but a clearer understanding of the Phase I to Phase II process is imperative to understand more fully the transition from sources of knowledge to implementation of knowledge. Also, we described trends in firm ownership by gender and by minority status, and the relationship between gender and minority status on outputs from Phase II projects. Case studies might illuminate the success of a firm receiving a Phase I and a Phase II award over time by these demographic characteristics.

To the best of our knowledge, researchers have not followed completed Phase II projects over time. The NRC datasets identified projects that commercialized, but since the names of the firms are masked in the datasets there is not a way to identify the successfulness of the commercialized technologies or to follow the growth trajectory of the firms. Such longitudinal studies could be accomplished through case studies or through survey methods.

These are just some of the roads that have been less traveled.

References

Allen, Stuart D., Stephen K. Layson, and Albert N. Link (2012). "Public Gains from Entrepreneurial Research: Inferences about the Economic Value of Public Support of the Small Business Innovation Research Program," *Research Evaluation*, 21: 105–12.

Andersen, Martin S., Jeremy W. Bray, and Albert N. Link (2017). "On the Failure of Scientific Research: An Analysis of SBIR Projects Funded by the U.S. National Institutes of Health," *Scientometrics*, 112: 431–42.

Antonelli, Cristiano and Claudio Fassio (2014). "The Heterogeneity of Knowledge and the Academic Mode of Knowledge Governance: Italian Evidence in the First Part of the 20th Century," *Science and Public Policy*, 41: 15–28.

Arrow, Kenneth J. (1962). "Economic Welfare and the Allocation of Resources for Invention." In *The Rate and Direction of Inventive Activity: Economic and Social Factors* (pp. 609–25), Princeton, NJ: Princeton University Press.

Audretsch, David B., Donald F. Kuratko, and Albert N. Link (2016). "Dynamic Entrepreneurship and Technology-based Innovation," *Journal of Evolutionary Economics*, 26: 603–20.

Audretsch, David B., Dennis P. Leyden, and Albert N. Link (2012). "Universities as Research Partners in Publicly Supported Entrepreneurial Firms," *Economics of Innovation and New Technology*, 21: 529–45.

Audretsch, David B., Dennis P. Leyden, and Albert N. Link (2013). "Regional Appropriation of University Based Knowledge and Technology for Economic Development," *Economic Development Quarterly*, 27: 56–61.

Audretsch, David B. and Albert N. Link (2018). "Innovation Capital," *Journal of Technology Transfer*, 43: 1760–7.

Audretsch, David B. and Albert N. Link (2019a). "The Fountain of Knowledge: An Epistemological Perspective on the Growth of U.S. SBIR-unded Firms," *International Entrepreneurship and Management Journal*, 15: 1103–13.

Audretsch, David B. and Albert N. Link (2019b). *Sources of Knowledge and Entrepreneurial Behavior*, Toronto: University of Toronto Press.

Audretsch, David B., Albert N. Link, and John T. Scott (2002). "Public/Private Technology Partnerships: Evaluating SBIR-supported Research," *Research Policy*, 31: 145–58.

Audretsch, David B., Albert N. Link, and Martijn van Hasselt (2019). "Knowledge Begets Knowledge: University Knowledge Spillovers and the Output of Scientific Papers from U.S. Small Business Innovation Research (SBIR) Projects," *Scientometrics*, 121: 1367–83.

Bastiat, Frédéric (1995, original 1848). *Selected Essays on Political Economy*, translated by Seymour Cain and edited by George B. de Huszar, Irvington-on-Hudson: Foundation for Economic Education. Available at:

https://oll.libertyfund.org/titles/bastiat-selected-essays-on-political-economy (accessed March 29, 2022).

Bearse, Peter M. and Albert N. Link (2010). "Economic Implications of Raising the Threshold Funding Limits on US Small Business Innovation Research Awards," *Science and Public Policy*, 37: 713–35.

Becker, Gary S. (1975). *Human Capital: A Theoretical and Empirical Analysis*, Chicago, IL: University of Chicago Press.

Bednar, Steven, Dora Gicheva, and Albert N. Link (2021). "Innovative Activity and Gender Dynamics," *Small Business Economics*, 56: 1591–9.

Birch, David L. (1979). *The Job Creation Process*, Cambridge, MA: MIT Program on Neighborhood and Regional Change.

Birch, David L. (1981). "Who Creates Jobs?" *Public Interest*, 65: 3–14.

Bozeman, Barry and Albert N. Link (2015). "Toward an Assessment of Impacts from U.S. Technology and Innovation Policies," *Science and Public Policy*, 42: 369–76.

Bush, Vannevar (1945). *Science—the Endless Frontier*, Washington, DC: Office of Scientific Research and Development.

Cunningham, James A. and Albert N. Link (2021a). "Latent Technology as an Outcome of R&D," *Technological Forecasting and Social Change*, 162: 120371.

Cunningham, James A. and Albert N. Link (2021b). *Technology and Innovation Policy: An International Perspective*, Cheltenham, UK and Northampton, MA, USA: Edward Elgar Publishing.

Dolfsma, Wilfred (2011). "Government Failure—Four Types," *Journal of Economic Issues*, 3: 593–604.

European Commission (2020). *Knowledge Transfer Metrics: Towards a European-wide Set of Harmonised Indicators*, Luxembourg: Publications Office of the European Union.

Foray, Dominique (2004). *Economics of Knowledge*, Cambridge, MA: MIT Press.

Gallo, Marcy E. (2021). "Small Business Research Programs: SBIR and STTR," Congressional Research Service report R43695, Washington, DC.

Gicheva, Dora and Albert N. Link (2013). "Leveraging Entrepreneurship through Private Investments: Does Gender Matter?" *Small Business Economics*, 40: 199–210.

Gicheva, Dora and Albert N. Link (2015). "The Gender Gap in Federal and Private Support for Entrepreneurship," *Small Business Economics*, 45: 729–33.

Gicheva, Dora and Albert N. Link (2016). "On the Economic Performance of Nascent Entrepreneurs," *European Economic Review*, 86: 109–17.

Granovetter, Mark S. (1973). "The Strength of Weak Ties," *The American Journal of Sociology*, 78: 1360–80.

Griliches, Zvi (1979). "Issues in Assessing the Contribution of Research and Development to Productivity Growth," *The Bell Journal of Economics*, 10: 92–116.

Hanifan, Lyda J. (1916). "The Rural School Community Center," *Annals of the American Academy of Political and Social Science*, 67: 130–8.

Hardin, John W., David J. Kaiser, and Albert N. Link (2020). "Public Support of Private Innovation," *Annals of Science and Technology Policy*, 4: 1–81.

Hayter, Christopher S. and Albert N. Link (2018). "Why Do Knowledge-intensive Entrepreneurial Firms Publish Their Innovative Ideas?" *Academy of Management Perspectives*, 32: 141–55.

Hayward, Mathew, Dean A. Shepherd, and Dale Griffin (2006). "A Hubris Theory of Entrepreneurship," *Management Science*, 52: 160–72.

Hess, Charlotte and Elinor Ostrom (2007). "Introduction." In Charlotte Hess and Elinor Ostrom, editors, *Understanding Knowledge as a Commons: From Theory to Practice* (pp. 3–26), Cambridge, MA: MIT Press.

Hicks, John R. (1932). *The Theory of Wages*, New York: Macmillan.

Hume, David (2007, original 1748). *An Enquiry Concerning Human Understanding* (edited by P.F. Millican), New York: Oxford University Press.

Humphrey, Thomas M. (1997). "Algebraic Production Functions and Their Uses before Cobb-Douglas," *Economic Quarterly*, 83: 51–83.

Joint Hearings (1979). "Joint Hearings before the U.S. Senate Committee on Commerce, Science, and Transportation and the Select Committee on Small Business; and to the U.S. House of Representatives Committee on Science and Technology and the Committee on Small Business," Washington, DC: Government Printing Office.

Lazear, Edward P. (2005). "Entrepreneurship," *Journal of Labor Economics*, 23: 649–80.

Le Grand, Julian (1991). "The Theory of Government Failure," *British Journal of Political Science*, 21: 423–42.

Leyden, Dennis Patrick and Albert N. Link (2015a). *Public Sector Entrepreneurship: U.S. Technology and Innovation Policy*, New York: Oxford University Press.

Leyden, Dennis Patrick and Albert N. Link (2015b). "Toward a Theory of the Entrepreneurial Process," *Small Business Economics*, 44: 475–84.

Link, Albert N. (2013). *Public Support of Innovation in Entrepreneurial Firms*, Cheltenham, UK and Northampton, MA, USA: Edward Elgar Publishing.

Link, Albert N. (2015). "Capturing Knowledge: Private Gains and Public Gains from University Research Partnerships," *Foundations and Trends in Entrepreneurship*, 11: 139–206.

Link, Albert N. (2021). "Knowledge Transfers from Federally Supported R&D," *International Entrepreneurship and Management Journal*, 17: 249–60.

Link, Albert N. (2022). "Vannevar Bush: A Public Sector Entrepreneur," *Foundations and Trends in Entrepreneurship*, 18: 1–74.

Link, Albert N. (forthcoming 2023). "The U.S. Small Business Technology Transfer (STTR) Program: An Assessment and an Evaluation of the Program," *Annals of Science and Technology Policy*.

Link, Albert N. and James A. Cunningham (2021). *Advanced Introduction to Technology Policy*, Cheltenham, UK and Northampton, MA, USA: Edward Elgar Publishing.

Link, Albert N. and Laura T.R. Morrison (2019). *Innovative Activity in Minority-owned and Women-owned Business: Evidence from the U.S. Small*

Business Innovation Research Program, Cham, Switzerland: Springer Nature Publishers.

Link, Albert N. and Christopher J. Ruhm (2009). "Bringing Science to Market: Commercialization from NIH SBIR Awards," *Economics of Innovation and New Technology*, 18: 381–402.

Link, Albert N. and Christopher J. Ruhm (2011). "Public Knowledge, Private Knowledge: The Intellectual Capital of Entrepreneurs," *Small Business Economics*, 36: 1–14.

Link, Albert N. and John T. Scott (1998). *Public Accountability: Evaluating Technology-based Institutions*, New York: Springer.

Link, Albert N. and John T. Scott (2009). "Private Investor Participation and Commercialization Rates for Government Sponsored Research and Development: Would a Prediction Market Improve the Performance of the SBIR Programme?" *Economica*, 76: 264–81.

Link, Albert N. and John T. Scott (2010). "Government as Entrepreneur: Evaluating the Commercialization Success of SBIR Projects," *Research Policy*, 39: 589–601.

Link, Albert N. and John T. Scott (2012a). "Employment Growth from Public Support of Innovation in Small Firms," *Economics of Innovation and New Technology*, 21: 655–78.

Link, Albert N. and John T. Scott (2012b). *Employment Growth from Public Support of Innovation in Small Firms*, Kalamazoo, MI: W.E. Upjohn Institute for Employment Research.

Link, Albert N. and John T. Scott (2012c). "Employment Growth from the Small Business Innovation Research Program," *Small Business Economics*, 39: 265–87.

Link, Albert N. and John T. Scott (2012d). "The Exploitation of Publicly Funded Technology," *Journal of Technology Transfer*, 37: 375–83.

Link, Albert N. and John T. Scott (2013). *Bending the Arc of Innovation: Public Support of R&D in Small, Entrepreneurial Firms*, New York: Palgrave Macmillan.

Link, Albert N. and John T. Scott (2018a). "Propensity to Patent and Firm Size for Small R&D Intensive Firms," *Review of Industrial Organization*, 52: 561–87.

Link, Albert N. and John T. Scott (2018b). "Toward an Assessment of the US Small Business Innovation Research Program at the National Institutes of Health," *Science and Public Policy*, 45: 83–91.

Link, Albert N. and John T. Scott (2020). "Creativity-enhancing Technological Change in the Production of Scientific Knowledge," *Economics of Innovation and New Technology*, 29: 489–500.

Link, Albert N. and John T. Scott (2021). "Technological Change in the Production of New Scientific Knowledge: A Second Look," *Economics of Innovation and New Technology*, 30: 371–81.

Link, Albert N. and Donald S. Siegel (2003). *Technological Change and Economic Performance*, New York: Routledge.

Link, Albert N. and Martijn van Hasselt (2020). "Exploring the Impact of R&D on Patenting Activity in Small Women-owned and Minority-owned Entrepreneurial Firms," *Small Business Economics*, 54: 1061–6.

Link, Albert N. and Mike Wright (2015). "On the Failure of R&D Projects," *IEEE Transactions on Engineering Management*, 62: 442–8.

Link, Albert N., Robert S. Danziger, and John T. Scott (2018). "Is the Bayh-Dole Act Stifling Biomedical Innovation?" *ISSUES in Science and Technology*, Winter: 33–5.

Link, Albert N., Christopher A. Swann, and Martijn van Hasselt (2022). "An Assessment of the U.S. Small Business Innovation Research (SBIR) Program: A Study of Project Failure," *Science and Public Policy*, DOI: 10.1093/scipol/scac049.

Link, Albert N., Martijn van Hasselt, and Silvio Vismara (2021). "Going Public with Public Money," *Small Business Economics*, 57: 1419–26.

Locke, John (1996, original 1690). *An Essay Concerning Human Understanding* (edited by K.P. Winkler), Cambridge: Hackett Publishing Company.

Machlup, Fritz (1962). *The Production and Distribution of Knowledge in the United States*, Princeton, NJ: Princeton University Press.

Machlup, Fritz (1980). *Knowledge and Knowledge Production*, Princeton, NJ: Princeton University Press.

Merton, Robert K. (1936). "The Unanticipated Consequences of Purposive Social Action," *American Sociological Review*, 1: 894–904.

National Academies of Sciences, Engineering, and Medicine (National Academies) (2016). *STTR: An Assessment of the Small Business Technology Transfer Program*, Washington, DC: The National Academies Press.

National Research Council (NRC) (2008). *An Assessment of the SBIR Program*, Washington, DC: The National Academies Press.

Nixon, Phil, Megan Harrington, and David Parker (2012). "Leadership Performance Is Significant to Project Success or Failure: A Critical Analysis," *International Journal of Productivity and Performance Management*, 61: 204–16.

Pinto, Jeffrey K. and Samuel J. Mantel, Jr. (1990). "The Causes of Project Failure," *IEEE Transactions on Engineering Management*, 37: 269–76.

Polanyi, Michael (1966). *The Tacit Dimension*, Garden City, NY: Anchor Books.

Polanyi, Michael (1974, original 1958). *Personal Knowledge: Towards a Post-critical Philosophy* (corrected edition). Chicago, IL; London: University of Chicago Press.

President's Council of Advisors on Science and Technology (PCAST) (2008). *University–Private Sector Research Partnerships in the Innovation Ecosystem*, Washington, DC: The White House.

Schultz, Theodore W. (1975). "The Value of the Ability to Deal with Disequilibria," *Journal of Economic Literature*, 13: 827–46.

Scott, Troy J., John T. Scott, and Albert N. Link (2017). "Commercial Complexity and Entrepreneurial Finance," *Economics of Innovation and New Technology*, 26: 489–500.

Shackle, G.L.S. (1979). *Imagination and the Nature of Choice*, Edinburgh, Scotland: Edinburgh University Press.

Shepherd, Dean A. and Johan Wiklund (2006). "Successes and Failures at Research on Business Failure and Learning from It," *Foundations and Trends in Entrepreneurship*, 2(1): 1–35.

Solow, Robert M. (1957). "Technical Change and the Aggregate Production Function," *The Review of Economics and Statistics*, 39: 312–20.

Terleckyj, Nestor E. (1974). *The Effects of R&D on the Productivity Growth of Industries: An Exploratory Study*, Washington, DC: National Planning Association.

U.S. Government Accountability Office (GAO) (1985). *Managing the Cost of Government: Building an Effective Financial Management Structure*, Washington, DC: U.S. Government Accountability Office.

Winston, Clifford (2006). *Government Failure versus Market Failure: Microeconomics Policy Research and Government Performance*, Washington, DC: Brookings Institution.

Wolf, Charles Jr. (1988). *Markets or Governments: Choosing between Imperfect Alternatives*, Cambridge, MA: MIT Press.

Index

accountability 112
Allen, Stuart D. 34
An Enquiry Concerning Human Understanding (Hume) 80
An Essay Concerning Human Understanding (Locke) 78–80
Andersen, Martin S. 35, 58
Antonelli, Cristiano 81
arrival curve 26, 27
assumptions 99
Audretsch, David B. 33, 34, 35, 36
Australia 26, 27

Bastiat, Frédéric 100
Bayh-Dole Act (1980) 24–5, 103–4
Bearse, Peter M. 33
Becker, Gary S. 58
Bednar, Steven 37
Birch, David L. 2
Bray, Jeremy W. 35
Budget and Accounting Act (1921) 112
Bush, Vannevar 9–10

Carter, Jimmy 10, 22
commercialization of technologies 4–5, 6, 13, 86, 122–3
    counterfactual analysis 118–19
    defining commercialization 89
    financial stakeholders 88–93
    market for SBIR developed technologies 94–7
    NRC datasets 31, 33–7
    productive capital 46
    project failure 54, 62
    technology policy 23, 24
    unanticipated consequences 101, 103, 104–10
consumer surplus 5, 34

counterfactual analysis 6, 112–19
Cunningham, James A. 26, 37

DHS *see* Department of Homeland Security
Department of Agriculture (USDA) 15
Department of Commerce (DOC) 15
Department of Defense (DOD) 15, 121
    counterfactual analysis 113, 114, 117, 118–19
    demographics of SBIR awardees 40, 43
    financial stakeholders 88, 89, 90, 93
    knowledge production functions 82, 85
    market for SBIR developed technologies 95
    NRC datasets 31–7
    productive capital 48, 49, 51–2
    project failure 55, 59, 62, 64, 65
    trends in patents received 104, 105
    unanticipated consequences 104, 105
    university collaboration 71, 72, 75
Department of Education (ED) 15
Department of Energy (DOE) 15, 121
    counterfactual analysis 113, 116, 117, 118–19
    Demographics of SBIR awardees 42, 43
    financial stakeholders 88, 89, 92
    knowledge production functions 84, 85
    market for SBIR developed technologies 95, 96
    NRC datasets 31, 32, 34–7
    productive capital 48, 50, 51–2

project failure 55, 57, 59, 61, 62, 64, 68
    trends in patents received 104, 108
    unanticipated consequences 104, 108
    university collaboration 71, 73, 75–6
Department of Health and Human Services (HHS) 15
Department of Homeland Security (DHS) 15
disadvantaged persons 11, 12, 14, 15, 39, 43
DOC *see* Department of Commerce
DOD *see* Department of Defense
DOE *see* Department of Energy
Dolfsma, Wilfred 117
Domestic Policy Review (Carter administration, 1979) 10, 22
DOT (Department of Transportation) 15
due diligence 6, 90, 103, 123

Economic Recovery Tax Act (1981) 24–5
ED *see* Department of Education
EPA *see* Environmental Protection Agency
Environmental Protection Agency (EPA) 15
experiential capital (EC) 44, 45, 46, 47, 48–52, 121
    counterfactual analysis 114–117
    knowledge production functions 81–5
    project failure 58–62

Fassio, Claudio 81
financial stakeholders 5, 88–93
    commercialization 88–93
    sales of process(es), product(s) and service(s) 89, 94, 95–7, 122
Foray, Dominique 81
'fountains of knowledge' 79–80

Gallo, Marcy E. 13–14
Gicheva, Dora 34, 35, 37
Government Accountability Office (GAO) 112
government failure 117–18
Granovetter, Mark S. 46
Griliches, Zvi 44

Hanifan, Lyda J. 46
Hayter, Christopher S. 36
HHS *see* Department of Health and Human Services 15
Hicks, John R. 44
human capital (HC) 44–6, 47–52, 121
    counterfactual analysis 114–17
    knowledge production functions 81–5
    NRC datasets 32–3, 35, 36
    project failure 58–62, 68
Hume, David 80
Humphrey, Thomas M. 44
hurdle rate 16–18, 101, 102–3

impressions 80
Initial Public Offering (IPO) 37
innovation policy, defining 23, 24
innovative outputs (IO) 44–5, 81, 85, 97, 101
international SBIR programs 4, 6, 26–7, 78, 122

Japan 26, 27

knowledge production functions
    contemporary perspectives 80–81
    historical perspectives 78–80
    patents 5, 78, 81–6
    productive capital 44–5, 78, 82–6
    scientific publications 5, 78, 81–6, 122
    social benefits of SBIR 78, 85, 86
knowledge transfers 5, 6, 26, 86, 94–5, 101, 110, 122
Kuratko, Donald F. 35

Layson, Stephen K. 34

Lazear, Edward P. 58
legislation
    demographics of SBIR awardees
        39–40
    legislative history of SBIR 3,
        8–19, 39, 121
    unanticipated consequences
        99–101, 110
Leyden, Dennis Patrick 25, 34
Link, Albert N. 16, 17, 25, 26, 33, 34,
    35, 36, 37, 58, 103–4
Locke, John 78–80

Machlup, Fritz 81
Managing the Cost of Government
    (GAO, 1985) 112
market failure 16, 17, 18
market for SBIR developed
    technologies 94–7
Massachusetts Institute of Technology
    (MIT) 2
Merton, Robert K. 100–101
minority-owned firms 4, 11, 12, 36,
    123
    counterfactual analysis 114–17
    demographics of SBIR 39–43
    productive capital 47–52
    project failure 59–62
Morrison, T. R. 36

NASA (National Aeronautics and
    Space Administration) 15, 121
    counterfactual analysis 113, 115,
        118–19
    demographics of SBIR awardees
        41, 43
    financial stakeholders 88, 89, 91
    knowledge production functions
        83, 85
    market for SBIR developed
        technologies 95, 96
    NRC datasets 31, 32, 34–7
    productive capital 48, 50, 51–2
    project failure 55, 56, 60, 64, 67
    trends in patents received 104, 107
    unanticipated consequences 104, 107
    university collaboration 71, 73, 76

National Academies (US National
    Academies of Sciences,
    Engineering, and Medicine) 2,
    30, 31
National Cooperative Research Act
    (1984) 24–5
National Defense Authorization Acts
    13, 31
National Institutes of Health (NIH) 121
    counterfactual analysis 113, 115,
        117, 118–19
    demographics of SBIR awardees
        41, 43
    financial stakeholders 89, 91, 93
    knowledge production functions
        83, 85
    market for SBIR developed
        technologies 95, 96
    NRC datasets 31, 32, 34–7
    productive capital 48, 49, 51–2
    project failure 55, 56, 60, 62, 64,
        66
    trends in patents received 104,
        106
    unanticipated consequences 104,
        106
    university collaboration 70–71,
        72, 76
National Research Council (NRC)
    datasets 2–3, 6, 30–33, 121, 123
    empirical studies 33–7
    patents 34, 36, 36
    productive capital 45–7
    project failure 54, 63, 69
    survey questions 45–6, 63, 70, 89,
        94–5, 113, 118
    university collaboration 34, 36,
        37, 70
National Science Foundation (NSF)
    15, 121
    counterfactual analysis 113, 116,
        118–19
    demographics of SBIR awardees
        42, 43
    financial stakeholders 88, 89, 92
    knowledge production functions
        84, 85

market for SBIR developed
technologies 97
NRC datasets 31, 32, 34–7
productive capital 48, 51
project failure 55, 57, 61, 62,
64, 69
trends in patents received 104, 109
unanticipated consequences 104,
109
university collaboration 70–71,
74, 75, 77
Neighborhood and Regional Change
report (MIT, 1979) 2
Netherlands 26, 27
New Zealand 26, 27
'newness' 58
NIH *see* National Institutes of Health
NRC datasets *see* National Research
Council (NRC) datasets
NSF *see* National Science Foundation

overconfidence 58

patents 5–6, 122
knowledge production functions
5, 78, 81–6
NRC datasets 34, 36, 36
unanticipated consequences
103–10
perceptions 79–80
performance metrics 4
Phase I projects 13, 40, 54, 101, 103
Phase II projects 4, 5, 13–14
counterfactual analysis 112,
113–19
demographics of awardees 39–42
failure of *see* project failure
financial stakeholders 88–93
knowledge production functions
78, 82, 86
market for SBIR developed
technologies 94–7
NRC datasets 27, 31–7
productive capital 45–7, 51
unanticipated consequences 99,
101, 103–10
university collaboration 70–77

Phase III projects 13
Polanyi, Michael 81
policy transfer 4, 6, 26–7, 78
President's Council of Advisors
on Science and Technology
(PCAST) 74–5
production functions 43, 44, 78–86
productive capital 4, 5, 43, 121
elements of 44–7
knowledge production functions
44–5, 78, 82–6
minority-owned firms 47–52
project failure 58–61, 69
quantifying elements 47–52
university collaboration 70, 75–7, 78
women-owned firms 47–52
see also experiential capital;
human capital; social
capital; technical capital
productivity slowdown 2, 8–10, 22
program assessment 112–13
program evaluation 112–13
program performance, understanding
112–13
project failure 4
empirical findings 58–62
measure of 54–8
productive capital 58–61, 69
reasons for 58, 63–8
university collaboration 75–7
public-sector entrepreneurship 25–7

R&D (research and development)
investment
financial stakeholders 88, 90
knowledge production functions 85
NRC datasets 35–6
private sector 9–10, 16–18
productive capital 45, 47
public-sector 16, 22, 23, 101–2
risk and uncertainty in 16–18, 26
unanticipated consequences 101–2
university collaboration 74
R&E (research and experimentation) 24
rate of return 16–18, 17, 101–3
reflections 78–81
risk and uncertainty in 16–18, 26

Ruhm, Christopher J. 33, 34

sales of process(es) 89, 94, 95–7, 122
sales of product(s) 89, 94, 95–7, 122
sales of service(s) 89, 94, 95–7, 122
SBIR *see* Small Business Innovation
    Research program
Schultz, Theodore W. 80
science, defining 23, 24
Science—the Endless Frontier 9–10
scientific publications 5, 36, 78, 122
Scott, John T. 16, 17, 33, 34, 35, 36
sectors of sales 94–7
Selected Essays on Political Economy
    (Bastiat) 100
sensations 78–80
Small Business Act (1953) 10, 14
Small Business Innovation
    Development Act (1982) 2, 9,
    10–13, 24, 25, 30, 39, 122
Small Business Innovation Research
    (SBIR) program
  commercialization of technologies
    4–5, 6, 13, 23, 24, 31
  counterfactual analysis *see*
    counterfactual analysis
  databases used *see* NRC datasets
  demographics of awardees 4,
    39–43
  economics of 16–18
  as element of U.S. technology
    policy 22–8
  empirical perspective 3, 4–5,
    30–97, 121
  empirical studies 33–7
  established 2
  failure of projects *see* project
    failure
  financial stakeholders *see*
    financial stakeholders
  innovative nature of 26, 30–31
  institutional perspective 2–4,
    8–28, 121
  knowledge production functions
    see knowledge production
    functions

lack of information on research
    projects 2, 123
legislative history of 3, 8–19,
    39, 121
market for SBIR developed
    technologies *see* market
    for SBIR developed
    technologies
mission statement 14, 18–19
Phase funding structure 4, 5,
    13–14
Policy Directive 13–14
policy perspective 3, 5–6, 99–123
policy transfer to other countries
    4, 6, 26–7, 78, 122
productive capital *see* productive
    capital
program goals 11, 14, 19, 39–40,
    43, 46
public-sector entrepreneurship
    25–7
purpose statement of 11–12, 14,
    18–19, 39
R&D investment 9–10, 16–18, 22
Reauthorization Act (2000) 3, 30
social benefits of *see* social
    benefits of SBIR
trend analysis of awardees 40–43
unanticipated consequences
    see unanticipated
    consequences
university collaboration *see*
    university collaboration
Small Business Research and
    Development Enhancement Act
    (1992) 30, 39
social benefits of SBIR 2, 4–6, 18, 26,
    27, 33, 121–3
  counterfactual analysis 113, 119
  financial stakeholders 93, 94, 97
  knowledge production functions
    78, 85, 86
  unanticipated consequences 99,
    104, 110
social capital (SC) 5, 44–5, 46–7,
    48–52, 121, 123
  counterfactual analysis 114–17

financial stakeholders 88, 90, 93
knowledge production functions
81–5
project failure 58–62, 68
South Africa 26, 27, 122
South Korea 26, 27
status quo economic environment 25–6
Stevenson-Wydler Act (1980) 24–5
survey questions 45–6, 63, 70, 89,
94–5, 113, 118

Taiwan 26, 27
technical capital (TC) 33, 44, 45
technology policy, defining 22–4
technology transfers 5, 6, 25, 26, 31,
74, 93, 95, 123
'The Unanticipated Consequences
of Purposive Social Action'
(Merton) 100–101
Turkey 26, 27

unanticipated consequences 5–6, 18,
99–110
legislation 99–101, 110
patents 103–10
rate of return 101–3
social benefits of SBIR 99, 104,
110
United Kingdom 26, 27
university capital (UC) 4, 70, 75–7,
81–5

university collaboration 4, 122
defining 70–71
knowledge production functions
81–5
NRC datasets 34, 36, 37
project failure 75–7
trends in 71–5
unanticipated consequences
103–4
university capital 4, 70, 75–7,
81–5
*University-Private Sector Research
Partnerships in the Innovation
Ecosystem* (PCAST, 2008)
74–5
USDA *see* Department of Agriculture

Van Hasselt, Martijn 36, 37
Vismara, Silvio 37

Wiklund, Johan 58
Wilson, Woodrow 112
Wolf, Charles Jr. 117
women-owned firms 4, 12, 14–15, 123
demographics of SBIR awardees
39–43
counterfactual analysis 114–17
NRC datasets 36, 37
productive capital 47–52
project failure 59–62
Wright, Mike 35, 58